*A History of
the Fish Hook*

Hans Jørgen Hurum

A History of the Fish Hook

and the story of Mustad, the hook maker

Adam & Charles Black
London

Norwegian edition first published 1976
by Grøndahl & Søn, Munkedamsvn 35, Oslo, Norway.

English language edition first published 1977
by A & C Black Ltd, 35 Bedford Row, London, WC1R 4JH
© Grøndahl & Søn Forlag A/S, 1976, 1977

ISBN 0 7136 1834 5

Filmset by Keyspools Ltd, Golborne, Lancashire.
Printed in Great Britain by
The Pitman Press, Bath

Contents

Does the fish with a hook in its mouth feel pain? This question has been debated time and again. Before identifying too closely with the fish we should remember that it is quite possible that the mouth of a fish is exceedingly tough and insensitive. For example, a wolf-fish loves sea-urchins and the cod regards live crabs as a delicacy. (Photo: Nils Viker)

Foreword

This book is an anniversary publication, commissioned by O. Mustad & Sön A/S, the famous Norwegian hook manufacturers.

However this is no ordinary company history – the reader will look in vain for the names of the great many people who would have to be mentioned in such a book. Instead, the author deals mainly with the fish hook – that unpretentious implement which has accompanied Man since the beginning of time. There are thousands of books about fishing and hunting, but this is probably the first time a general book on fish hooks has been written.

Many illustrations of fish hooks in this book could have been reproduced photographically. However, to ensure clarity of detail they have been drawn with the most scrupulous accuracy at the Mustad design office.

For invaluable assistance at home and abroad the author is indebted to so many persons that it is impossible to name them all. A special word of thanks goes to Dr Andres v. Brandt, the former director of the Institute of Fisheries Research in Hamburg, for his marked interest and willingness to assist during these three years. I am also greatly indebted to the directors and personnel of the Mustad company, who never tired of explaining and expounding. Even to those who work with it day in and day out, the fish hook remains an engrossing subject.

Oslo/Ula 1976

HANS JORGEN HURUM

7

The hook

There is a minimum of technical terminology in this book, but a few words need some explanation.

Most people know what is meant by the *eye* of a hook. However, in Mustad terminology you will find two words, 'ring' and 'eye' used, apparently meaning the same thing.

In fact, Mustad use the term 'ringed hooks' as a general description, retaining the description 'eyed hooks' to indicate a particular kind of hook where the ring is parallel to the bend.

'Point' is another ambiguous term. It can be the sharp point at the very end of the hook, which is its correct meaning when measuring the 'gap' (the distance from the point to the shank). But when one talks about a 'hollow point' or a 'superior point', one is describing that part from the bottom of the barb to the sharp end of the point. Having said that, it must be stated that 'gap' has no official place in Mustad terminology: they prefer to speak of the 'width of the bend'. In the same way, expressions like 'crook' or 'bite' (the distance from the bottom of the bend up to the sharp end of the point), are avoided: here they prefer to speak of 'the depth of the throat'.

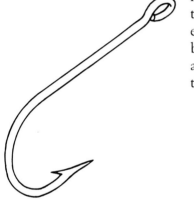

Example of a 'ringed' hook.

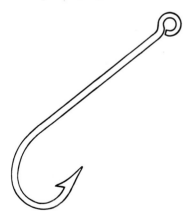

Example of an 'eyed' hook.

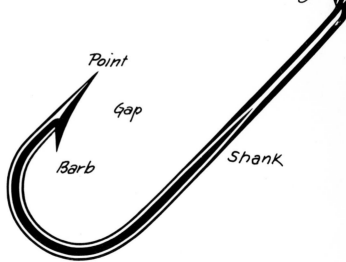

9

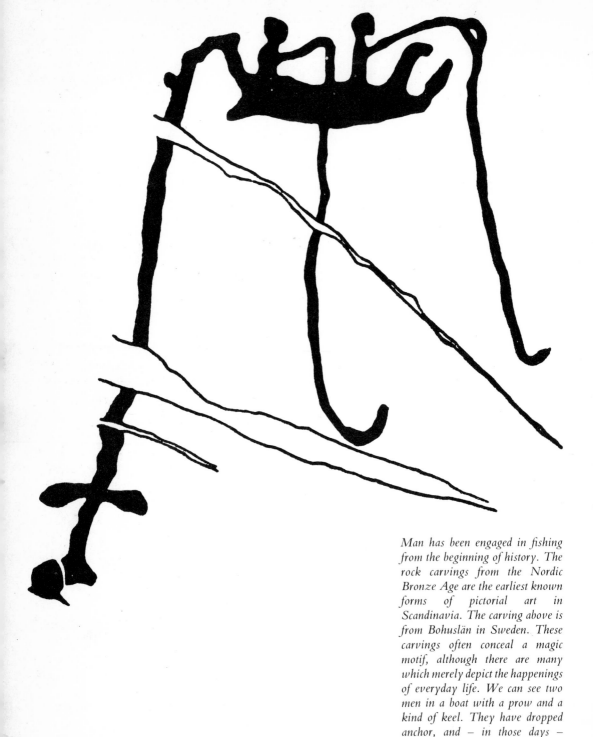

Man has been engaged in fishing from the beginning of history. The rock carvings from the Nordic Bronze Age are the earliest known forms of pictorial art in Scandinavia. The carving above is from Bohuslän in Sweden. These carvings often conceal a magic motif, although there are many which merely depict the happenings of everyday life. We can see two men in a boat with a prow and a kind of keel. They have dropped anchor, and – in those days – certainly wouldn't have had to wait very long for a bite.

From the Stone Age to Charles Kirby

In a little town in a little country – Gjøvik in the heart of Southern Norway – there is a factory which is like something out of a fairy-tale. By international standards, it is not a gigantic concern, but it has the whole world for its market. Mustad's fish hooks can be found practically anywhere on earth.

This book will try to describe the fascinating history of the fish hook, but first, a few examples of the remarkable way in which a factory located in a remote inland community has been able to develop into a market leader throughout the world.

Erik Holtedahl, a young Norwegian, was stationed at the Embassy in Moscow. One Sunday, while taking a walk, he came to a pond where an old man sat engrossed in his angling. According to Holtedahl, the man could tell by his accent that he was a foreigner and asked where he came from. 'Norway,' he replied. 'You make good fish hooks there,' said the man. He complained about the day's poor catch. Import restrictions had been imposed in the Soviet Union, and this annoyed the man, because he had far better catches with the Norwegian hooks.

In 1970, Robert Fossberg, a Norwegian employed in Canadian shipping, came across an unusual implement that Indian fishermen used at Lake Mistassini, east of James Bay. During the winter, when fishing on the ice, they let the line drift under the edge of the ice, where the bigger fish come up in their search for food. The Indians fasten a wooden stick to the hook in order that it will float and then wind a thong of leather around the stick and a good way up the leader. The equipment seemed strangely ancient and primitive but the hook itself, upon closer examination, turned out to be a Mustad hook.

And now from Canada to the Pacific: Pitcairn is one of the loneliest spots on earth. Living on this steep little island are descendants of the celebrated 'Bounty' mutineers, who, in the revolutionary year of 1789 rose up in arms on a British naval vessel, set the captain and his closest men adrift in a life boat, and took over the ship. Some of the mutineers went ashore on Pitcairn Island, which was uninhabited. Here they burned the ship so that no trace of the work of man was visible from the sea. After nine years only two of the mutineers remained alive. Smith was the name of one of the men, Young the other. They had brought women with them from Tahiti, and from these founders the community grew.

The Mustad factory has had the Pitcairn islanders as customers for a long time. The correspondence between Gjøvik and Pitcairn has been continuous, if not exactly regular.

From time to time, tourist ships called at Pitcairn, to

The following letter, dated 1961, was obviously written on borrowed stationery on board a tourist ship, and is addressed to the then export manager.

Dear Johannes Westerby, our Dear friend, many thanks for your many favour & kindness in helping us to get the money, and also in supplying the men with your extra strong Mustad fish hooks, King hooks of the world, the men who ordered the hooks are truly grateful with it. So I can only say a big THANK YOU *for your many kindness to us all on lonely Pitcairn. Hope this find you all well, I remain yours truly.*
Theodore Young.

have a look at the island and its singular inhabitants. Then the Pitcairn dwellers began trading with the tourists. They borrowed ship's stationery and ordered their wares – for example from O. Mustad & Søn, Gjøvik. Sometimes they would enclose pound notes in the envelope, but often they were short of money. Then they would gather shells and stones on the shore, make jewellery out of them, and ask if Mustad would trade the jewellery for fish hooks. The employees at the hook factory paid good prices for the jewellery, the money went in the firm's till – and the hooks were supplied to Pitcairn.

The hook factory in Gjøvik is only a hundred years old. But the hook itself – how old is that? No one knows. Some facts are known, but, to a certain extent, scientists are unsure about the way man lived in the far distant past. According to Dr Anatol Heintz, a Norwegian palaeontologist:

Man has very likely eaten fish from the moment that he became a human being, i.e. from the time he became a two-

This photo, originally a colour photo, (reproduced from J. G. Clark The Stone Age Hunters*) is from present-day Australia. Little imagination is needed to picture primitive man in a similar position at some time in the distant past.*

legged creature who left the forest and stayed in open savannahs and sought rivers and lakes. He no longer used his free hands for walking or climbing only, but, among other things, for making implements. His brain had already begun its miraculous development which, in the course of an amazingly short time, would turn him into the most intelligent creature on earth.

Not many years have elapsed since Dr L. S. B. Leakey, the British anthropologist, made a sensational discovery in the Olduvai Chasm, a deep crevasse in Tanzania, Central Africa. He found the skeletal remains of primitive people – an ape-like man who walked erectly (Australopithecus), and possibly true man (homo). Their settlements were excavated along the shores of a lake that has long since vanished. In addition to highly primitive stone implements, remains of animals were found on which these men and ape-like men subsisted: lizards, tortoises, snakes, small mammals and fish.

The most interesting aspect of these finds is the fact that scientists have been able to determine their age with considerable certainty. They are approximately 1.8 million years old.

Prehistoric methods of catching fish survived for a long time. As late as 1885 scenes like this could be witnessed in Samoa. People served as nets, and the catch was put to death by blows. (Natur-Photo Fritz Siedel, reproduced from v. Brandt, Das Grosse Buch vom Fischfang, Pinguin Verlag, Insbruk, 1975.)

But history does not tell us how they caught their fish. It may safely be said that it was not with hooks. Instead, it must be assumed that the first fishing equipment was arms and fingers. Thus, it would be

Since the dawn of time, fish have been caught with the gaff hook alone. Here is an illustration of ice fishing in the Urals. The Russian fishermen used long poles for gaffing the fish.

The poles, equipped with primitive hooks, are pulled haphazardly with short quick jerks. The fish in question are sturgeon and European catfish. The sturgeon is a curious fish which approaches the gaff of its own accord. As the picture indicates, this method of catching fish has persisted all the way down to the present time, especially at the estuaries of rivers emptying into the Caspian Sea. (J. Jánco, Dritte asiatische Forschungsreise des Grafen Eugen Zichy, Budapest, 1900.)

reasonable to believe that man gradually learnt to use a pole or a forked branch as an extension of his arms. If the pole had a sharp point he could spear the fish, or – who knows – jam the fish between the prongs of a fish spear, or perhaps hook them with some kind of a gaff.

Modern man (homo sapiens sapien), in the shape of Cro-Magnon Man is a late-comer to the scene – perhaps only thirty to forty thousands years ago. According to Professor Heintz: 'Very little is known about the origins of these people and how they came to Europe. But it may safely be assumed that they were familiar with fish hooks long before they came to Europe.'

Practically all of the known fish hooks from the Stone Age are to be found in museums. The number, of course, is small in proportion to their former distribution. The most obvious reason is the perishability of the materials. Wood rots, and even shells, bone and horn deteriorate with the years.

There is very good reason to believe that the oldest fish hooks were of wood. If one takes a branch with twigs that stick out at suitable angles, it would take very

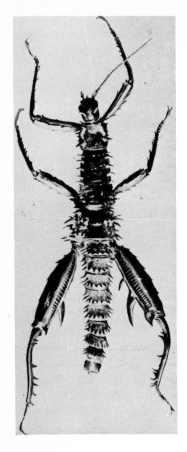

From New Guinea, a 'natural' fish hook made of an insect, Eurycantha latro. *The unusually strong leg of the male makes a ready-made hook. (Photo from W. Radcliffe,* Fishing from the earliest Times, *New York, 1969.*

little to make it into a reasonably good hook. Or think of the many plants that have natural thorns. Who could wish for a sharper point than the pointed thorns on a hawthorn bush. In the British Isles fishermen from Wales to the Thames have caught flounders with hawthorn hooks, usually with earthworms for bait. It matters little that the point of a hawthorn hook is without a barb – nowadays there are still plenty of barbless hooks on the market. Many kinds of natural hooks have been used – for example, among the Indians, the claw of a hawk and the beak of an eagle.

There is considerable evidence that wooden hooks can be strong enough for angling. In Africa as well as Asia, crocodiles have been caught with wooden hooks. Even today, along the coast of New Guinea, there are still fishermen who catch the shark with them. According to Dr Bengt Anell these hooks (more than half a metre in length) are probably the biggest fish hooks in use today.

Many people assume that the use of wooden hooks must have been more or less impractical. Since wood floats, the hook would probably have to be fastened to a stone or something else that was heavy enough to make it sink. But it would be a rash assertion to maintain that fish will not take a floating hook. The fact is that fishermen have often regarded floating hooks as an advantage. Up to the end of the nineteenth century, and perhaps even later, Lapp fishermen used wooden hooks in the great cod fisheries in Lofoten. They carved their hooks of juniper, a tough variety of wood, and burned the point to make it hard. The Lapplanders tried iron hooks but claimed that wooden hooks gave better catches because they seemed more lifelike in the water.

Wooden hooks are known throughout the world and in many places they are still in use, even in modern Sweden. In a book about fishing in Swedish lakes (J. G. Gyllenborg, *Kort Afhandling om Insjö-Fisket . . .* , Stockholm, 1770) one reads that burbot (lota-lota) are attracted by the smell of juniper. These wooden hooks or gorges were frequently twigs with three sharp points. Two hundred years later, in an article in *Svenska Dagbladet*, (3 March, 1963), old fishermen claimed that this special shape is the only one that is any good when it comes to catching them. The turbot spits out ordinary

Right: *A type of hook used by fishermen in Småland, and the method they use for fixing the 'hook'. Only one of the three points sticks out from the bait fish, and serves as a barb when the bait is swallowed. (Illustration from the Norwegian magazine* Fiskesport, *1957.)*

Below: *A bundle of leaders with juniper 'hooks' or 'gorges' for longline fishing in Swedish lakes (Nordiska Museet, Stockholm).*

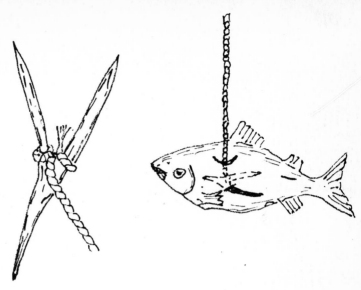

steel hooks, but – it is said – juniper hooks with three sharp points are impossible to dislodge.

In the Norwegian magazine *Fiskesport* (1957 no. 3) there is an eyewitness account of the way fishermen in Småland set their longlines for fish and eel with the same type of wooden hooks. The baited hooks float high enough above the bottom, so that snails and other creatures cannot get hold of the bait. Like the Lapplanders in Northern Norway they used juniper, but in Småland they did not burn the points, preferring to dry their hooks in the oven.

Among the many types of wooden hooks, the most surprising are those that the Indians of Alaska used for halibut. They are to be seen in a number of museums. Their shape is difficult to explain. Sometimes the shank was carved in the shape of a human being, a shaman or a god. But most often this type of hook has no decoration at all. The strange thing about these halibut hooks is the long point that is placed in a slanting position and extends so remarkably far towards the shank. The hook is made of two different kinds of wood, the point being an exceptionally hard variety.

Closely related to the halibut hooks from Alaska are the so-called ruvettus hooks from the southern Pacific (see page 80).

Hooks of stone and shell from the Stone Age have been discovered in widely different regions, although

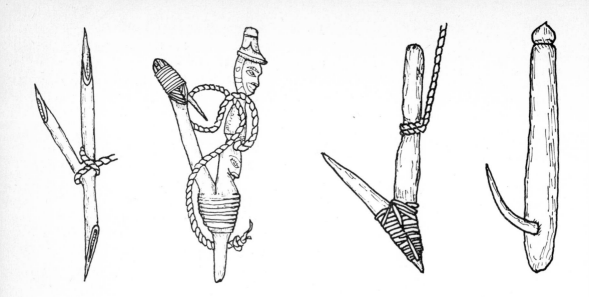

rarely in Europe. Unusually old hooks of shell have been found on the islands off Los Angeles and along the Pacific coast of Chile.

But the most typical hooks from the Stone Age – at least from the later palaeolithic age – were made of bone. When used for hooks, bone had a tendency to break; otherwise it withstood hard use, even in salt water. The processing required patience, but the Stone Age man had implements good enough to make extra fine hooks from this material.

The fact that no one knows when bone hooks came into use, is largely because bone seldom survives as a material. Only under exceptionally favourable conditions can bone be preserved for thousands of years. Extra-calcareous soil is needed. If there is not enough lime in the soil, the calcium in the bone is dissolved by the acids in the earth, and the bone will crumble to dust.

The oldest known bone hooks seem to be the ones that have turned up in Czechoslovakia during the excavation of the skeletal finds from late palaeolithic times. Dramatic scenes from the past are unfolding here. In all likelihood animals were killed mainly during organised, collective hunting. The animals were ringed in and killed with spears, harpoons, stones, arrows, or else they were driven into pitfalls or over the face of a cliff. This explains the enormous finds of the remains of animals in limited areas. One of the finds in Czecho-

slovakia – in Predmosti, Moravia – include the bones of more than a thousand Mammoths.

Throughout the warmer parts of Europe, one encounters gathering places like these with the remains of animals. In Solutré in France (Saône-et-Loire), the bones of herds of wild horses have been found; in the Ukraine, the bones of hundreds of bisons.

It is in such accumulations of bones, in soil containing lime, that the oldest traces of a Stone Age bone culture are to be found. According to J. Jelinek (*Das grosse Bilderlexikon des Menschen in der Vorzeit*, Artia, Praha, 1972) the fish hooks found in Moravia may be up to 20,000 years old, in other words from about the time when Stone Age artists carved and painted scenes of hunting and fishing in their dwellings – such as the famous caves in Spain and in Southern France.

Ancient bone hooks have also been found in Egypt and Palestine. The oldest, found in Palestine, is believed to be 9,000 years old.

Strangely enough, the Nordic countries are not far behind. The fact is that temperatures rose in astonishingly rapid cycles after the last Ice Age came to an end some 10,000 years ago. Over long periods the climate of Scandinavia was far milder than it is today. This made early settlement possible in the Nordic countries. Many valuable finds have been made in the so-called kitchen dunghills in settlements along the coast, where shellfish were an important part of the diet. Layer after layer of the shells of oysters, snails and mussels, as well as the bones of animals (including wild boar) birds and fish, have been found in calcareous soil. The oldest hooks from Maglemose (Sjaelland) are more than 8,000 years old. In Norway, the oldest known fish hooks were dug up in Visthulene in Jaeren. It is possible that they are older than previously assumed; some of them might approach the age of the Maglemose hooks.

The Norwegian finds of bone material, on a ledge called Skipshelleren near Bergen, are rather more recent. This is the richest discovery of bones that has been made in Norway, and among the wealth of implements here – tools and equipment for hunting and fishing – fish hooks have been found that show painstaking workmanship. A large discovery has also

19

been made in Kjelmøy in Finnmark. The hooks here are from a much later period, (the century around the birth of Christ), and it is a curious fact that the hooks were made of bone long after copper and iron were being used in other places. The hooks here are probably Lappish and reveal a culture that is clearly based on Stone Age traditions.

In some regions, such as Easter Island and several other islands in the South Pacific, fish hooks of stone and bone were indispensable. As there were no large mammals on Easter Island, there was a shortage of bone and the custom was adopted of making hooks of human bone. This they had in plenty since human sacrifices were made on Easter Island until the first missionaries arived at the turn of the last century. However, Thor Heyerdahl, the Norwegian expert on Easter Island, dismisses as pure invention the allegations of certain writers that there was any direct connection between a need for bone and the sacrifices. The use of this material is well known in other areas where there is no association with sacrificial ceremonies.

The future may bring surprising new discoveries. It is a fact that the last Ice Age, which was world-wide, caused enormous geographic upheavals. Great heat waves followed in the wake of the Ice Age, and seas and land rose and sank at an uneven tempo. In the Oslo area, shortly after the Ice Age, the sea was more than 200 metres higher than it is today. Thus, in order to find the Stone Age hooks that may have been concealed by nature, archaeologists would have to search above and below existing shorelines. Hunters and fishermen lived in areas that are today covered by water. For thousands of years after the Ice Age, England was connected to the continent by land. Thus, it has happened more than once that trawlers from the North Sea have brought ashore strange objects gathered from the sea bed – weapons, hunting and fishing implements from the Stone Age.

Stone Age people were familiar with many fishing methods, and some of these have probably been of greater importance than fishing with hooks. In every latitude nets were used to capture animals as well as fish. In addition there were stones that could be thrown. The bow and arrow made its appearance. In Japan a sea

Animals and birds, like otters and cormorants, may be trained to bring fish to people. This cormorant has a ring around its neck to prevent it from swallowing. It keeps the fish in its beak and flies back to the 'fisherman' with the catch. Here it receives its share and soon learns that the method pays. The system has been widespread, but is largely kept alive in our time as a tourist attraction, especially in the Far East (James Hornell, Fishing in many Waters, *Cambridge, 1950).*

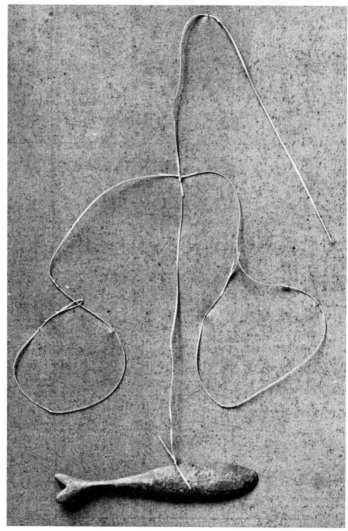

Right: Capturing fish in snares is far from unknown. The method seems to be designed for fish with powerful and stiff back fins. (Illustration from Nordiska Museet, Stockholm.)

bream from the Stone Age has been found that was shot through the head with an arrow of stone. But in this book we cannot go into the history of the countless different methods of catching fish – traps, nets, obstructions, poison, training birds and animals, snares and weirs – that have been invented by man.

It is strange to see how the Stone Age culture gave rise to the same main types of implements, regardless of how far away people lived from one another. Like fish hooks, harpoon heads of bone crop up in similar shapes from Greenland to Polynesia.

As a rule, the head of the harpoon was fastened to a pole and came away from the handle when it had

Great quantities of Stone Age harpoon heads have been found in many different areas. The material used was often the antlers of reindeer and stags. The shape and size may vary. Some of the harpoon heads have jagged edges on both sides, just as the number of barbs may vary considerably. The example shown here was found in a cave in the Basques, and is estimated to be some 12,000 years old. (Illustration from M. Almagro, Tratado-de Prehistoria, *Madrid 1959.)*

penetrated the animal or the fish. A rope was attached to the harpoon to make retrieval possible. The bone that was used had to be of the strongest type – reindeer horn or, preferably, stag's antlers. The extent of the finds, and their diversity, reveal how effective the equipment has been. In Finland, as well as in Sweden, the remains of seals have been found with harpoon heads of bone embedded in their skeletons.

It is not surprising that Stone Age harpoon heads have also served as fish hooks. In the Middle East (Armas Salonen, *Die Fischerei im Alten Mesopotamien*, Helsinki, 1970) it is known that an implement with several barbs was used as a kind of casting plug. The fisherman swung the line around his head a few times, to gather momentum, and then threw it out in the water. He did not use bait, but threw it where the fish had gathered, the idea being that in a shoal the implement would hook the bodies of fish at random.

A somewhat diffuse Biblical quotation may be relevant in this connection: it is from Matthew 17, 27, where the Bible, for once, mentions a hook and not a net. Some people believe that the use of the word hook here is due to a misunderstanding, that the meaning has been distorted in the translation. But if we imagine a hook like the oriental casting plug, the meaning makes some sense. The quotation is Jesus' reply to the question of whether he feels obliged to pay tribute money to the leviers of the Temple tax:

Notwithstanding, lest we should offend them, go thou to the sea, and cast an hook, and take up the fish that first cometh up: and when thou hast opened his mouth, thou shalt find a piece of money: that take, and give it unto them for me and thee.

But enough of harpoon heads and their possible variations: let us consider the even more interesting 'gorges'. Among several types of eel hooks, Mustad manufactures one that looks like this:

This special type, a completely straight hook, hardly qualifies to be called a hook. In principle, it corresponds in shape to a completely different fishing implement – the gorge. This is probably older than the hook. Some people believe that the gorge was the prototype for the fish hook, but, in all liklihood, it should be regarded as a parallel development.

22

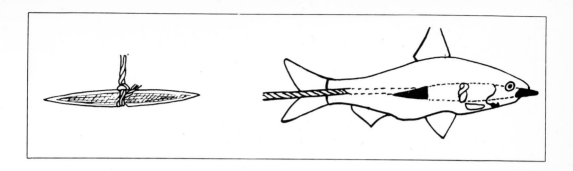

A simple gorge – and one of the ways it can be used in conjunction with live bait. In the picture of the bait fish, the black areas indicate the parts of the gorge that are visible on the outside of the fish. The dotted lines suggest the position of the leader and the gorge inside the bait fish. (Illustration from Stavanger Museum.)

In its simplest form, the gorge is merely a stick attached to the end of a line. The fish swallows the stick lengthwise; when the line is jerked the stick lodges crosswise and the fish may be hauled in. Here we are dealing with an extremely ancient fishing implement. It is well-known in practically every country and must have resulted in good catches, since it remained in use for thousands of years.

A number of years ago a Swedish scientist came to the Museum of Fisheries in Bergen to study the gorges that had been discovered in Norway. He said he would travel on to Finland from Bergen, because it had been rumoured that the gorge was still in use there. Sometime later, the Museum received a letter from this same scientist; it had not been necessary for him to go all the way to Finland, because it had turned out that the gorge was still widely used in his home district in Sweden.

Perhaps the people there were somewhat secretive about it; the gorge is often used illegally, in conjunction with live bait. At first, live bait was probably not used. Practically any stick can tempt a fish to bite, and this is especially true if the stick moves. Take a wooden peg that resembles a little fish, drag it through the water or suspend it from a string in running water, and sooner or later a larger fish will have a try at it. But it was soon discovered that a better catch resulted if a bait was attached to it, and anyone unscrupulous enough to use live bait would be guaranteed a catch.

The gorge can be made of any material, from wood and bone, to stone and metal, and it comes in all sizes. Not only Indians and Eskimos, but also Europeans have

23

What should we call this – a hook or a gorge?
What we see here is a Mustad 'hook' made of steel and sold on the market as 603 B.

used gorges for catching fish as well as fowl. The gorge does not attract fish alone. Birds will swallow a gorge baited with a little fish. In the Nile, and rivers in Thailand, people have captured crocodiles with them.

There were, of course, transitional shapes between the hook and the gorge. One example is the previously-mentioned wooden hook with three sharp points. Another example is a 'hook', made of the collarbone of ptarmigan, as used in Finland. A third example is a steel hook that Mustad makes, marked 603 B. Curiously, the only market for this is Greece.

Oddly enough, it is not so easy to define a fish hook. A hooking implement at the end of a pole ought to be a hook of sorts. But when we speak of a fish hook today we usually mean a hooking implement attached to a line. Human ingenuity is considerable, and hooks are not the only things that have been used by man in order to catch fish on a line.

If a bundle of cobwebs is attached to a line, the fish's teeth will become entangled in the web and it can be hauled up. This method is still in use in the Solomon Islands in the Pacific. In the Bosphorus, a certain species of tuna is fished in the same way, but using a bundle of silk. The Indians in California attached a bundle of vegetable fibre or human hair to the line sometimes with a live worm fastened to the bundle. Another method still used to catch eels in several countries, including Scandinavia, is a bunch of worms tied together with fine thread in which the eel's teeth get caught.

There are many other variations of this system. In classical literature the story is told of a man who fished for eels with the intestines of a sheep. When the eel had swallowed enough of the gut, the man would begin to blow with all his might so that the gut became too big for the eel to spit out again.

Looking at all the things that may be called fish hooks from the Stone Age, we can see that there are several different types.

When two pieces of bone were tied together the result resembled a wooden hook in which the twig stuck out from the branch at a more or less acute angle. But most characteristic of the bone hooks are the rounded forms.

The explanation may very well be that the bone

which was used often had a naturally rounded shape. It could be a knuckle, for example, or a leg bone that had been sawn crosswise. If so, work was saved by cutting as little of the ring-shaped material as possible. Perhaps this is the primitive explanation of the many, almost circular, bone hooks from the Stone Age in which the point extends inwards, remarkably close to the shank. Where the opening between the point and the shank is very small, the hooks may have been unfinished products. The people of the Stone Age may have had good catches with round hooks like these (almost circular hooks are available on the market to this very day). In addition, they may have learned from experience that round hooks are not easily snagged in seaweed, branches, stones and coral.

People of the Stone Age were able to bore holes as well as saw. If it was a piece of bone that was to be transformed into a hook, the boring of the hole was the first stage of the job. For stone hooks, too, the procedure was to start with the hole.

There is a great diversity of detail in the bone hooks. Grooves and barbs may be made at the top as well as the bottom. The gap is alternatively broad and narrow. A barb may also be found on the outer side of the bend, or a hole may be bored where it is least expected.

Everything must have had a purpose. The grooves at the top of the shank would have been useful in fastening the line. And if there were barbs under the bend, or on the outer side of the point, the explanation may often be that they were of use in attaching the bait.

A controversial phenomenon are the holes that were sometimes bored underneath the bend. These holes may have been used for fastening the bait, but there are scientists who are of the opinion that this type of hook was suspended upside down in the water. A piece of string was tied to the hole and then to a float; but more about this when we get on to sturgeon fishing.

As a rule it appears that the most ancient hooks were made without barb or any other refinement. The oldest hooks that have been found in Denmark and Norway indicate that only after thousands of years were they equipped regularly with barbs as well as grooves, bulges or holes to facilitate attachment of the bait and line.

Norwegian fish hooks from the Stone Age

Forty-three hooks and the remains of hooks have been found in the Vistehulene caves at Jaeren. The oldest are possibly 7,000 years old, primitively shaped like the one to the left.

The stone hook on the right is from Neolithic times, found at Sele in Klepp, Jaeren.

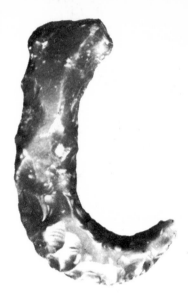

Below: Three types of hooks from the rich find at Skipshelleren, Straume in the Bergen district. The implements here are more painstakingly made. They include 120 bone fish hooks. The different layers of the find indicate that people were living here as cavemen for some 3,000 years. At that time Skipshelleren was close to the edge of the water; today it is situated 20 metres above sea level.

The four fish hooks below are from the large collection of bone implements that were found at Kjelmöy in the Varanger district of Finnmark. For the most part the objects here were made during the centuries immediately before and after the birth of Christ and probably bear witness to a Lapp-lander culture based on Stone Age traditions long after metals had begun to be used in more southerly regions. Reindeer horn was a familiar material to the Lapp-lander, and their fish hooks confirm this.

Above: *This bone hook from Meling in Håland seems to have been a good hook for fishing in Neolithic time.*

Below: *No one will dispute the beauty of this hook. It was found at Jortveit in Eide, Aust-Agder, and is considered to be 4,000 years old.*

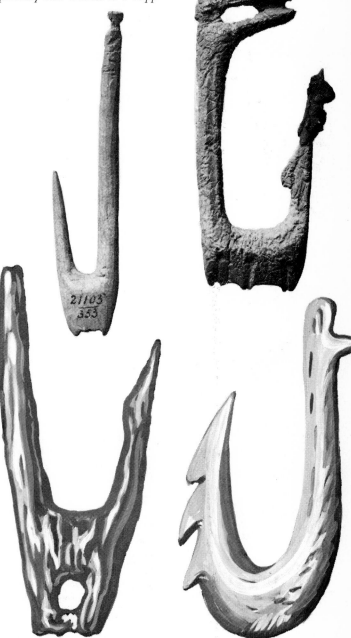

Fish hooks from all over the world

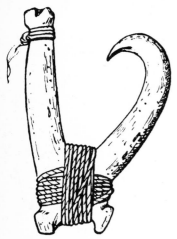

From Easter Island, probably made from human bone.

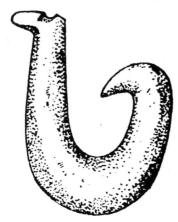

A stone hook from an early settlement on Pitcairn.

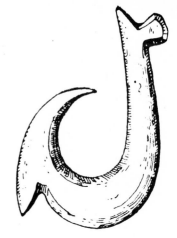

A hook of whale tooth, Hawaii.

A bone hook from Maglemose, Denmark, c. 6,200 BC.

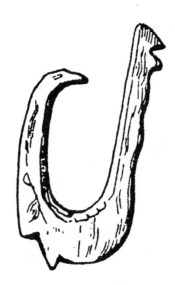

A Japanese hook of reindeer horn.

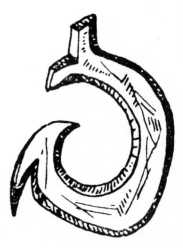

A hook of whale bone, Marshall Islands.

The illustrations on this page show both Stone Age hooks and hooks that have retained traditional shapes from ancient times. The drawings here are mostly from Bengt Anell, *Contribution to the History of Fishing in the Southern Seas*, Uppsala, 1955.

A compound hook from Volosova, Russia.

A shell hook from Santa Barbara, California.

A hook of tortoise bone, Ellice Islands.

A tortoise shell hook, Tobi, Micronesia.

A bone hook, from Vinda, Jugoslavia.

The fang of a bear, from Moosedorf.

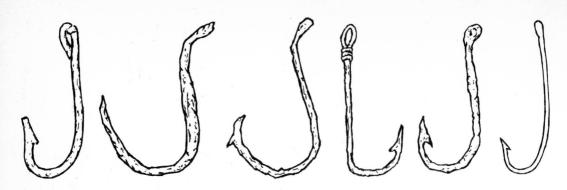

In the Middle East copper and bronze were used at a time when the countries of the north still found themselves in the Stone Age. From the left above: A copper hook from the Indus valley. Next two copper hooks from Mesopotamia, the oldest from 2600 BC, found in Ur (from Armas Salonen, Die Fischerei im alten Mesopotamien, The Academy of Science, Helsinki, 1970). Hook no. 4 is a bronze hook from a Rhodos grave at the time of Myceneaen civilization about 1400 BC (British Museum, London). No. 5 is a strong bronze hook found in a man's grave at Sande, Vestfold, Norway, from about 200 BC. Farthest to right a slender bronze hook from late in the Middle Ages, forged in the district around Mjösa (Hedmark Museum).

At first glance compound hooks look primitive. Their components, often of different materials, are tied together or joined in some other way. But probably the compound hooks were the strongest. While it is easy to break a slender, rounded bone hook, it would take a lot to break a securely tied compound hook.

It has been roughly estimated that copper came into use around 4000 BC followed by the gradual development of bronze. Among the oldest civilizations in which copper was utilized, were those along the banks of the Euphrates and the Tigris rivers, rivers abounding in fish and with enormous volumes of water.

Crete is renowned for especially rich finds of bronze hooks, followed closely by Italy. The bronze hooks that have been dug up at Pompeii and Herculaneum are beautiful and have a 'correct' shape. They are masterpieces of craftsmanship.

On the threshold of classical antiquity fish culture began to assume the forms we would recognize today. In ancient Mesopotamia the art of breeding fish in ponds was already known. The fish offered for sale were dried, salted or smoked. Commercial fisheries developed throughout the Middle East and in the Mediterranean countries. In the Golden Age of the Roman Empire, the trade in salted fish made up a considerable part of the ships' traffic in the bustling Roman port of Ostia. The epicures of Rome were experts at preparing fish

Left: *In ancient days fish hooks were used for barter and often as coins. Later regular coins would take the symbolic shape of a hook. The 'hook' here has never been used for fishing. Even after the* British had occupied Ceylon in 1815, 'hook silver' continued to be legal tender in Ceylon (*A. Hingston Quiggin*, A Survey of Primitive Money, *London, 1949*).

Right: This is considered to be the oldest picture of angling. The Egyptian fishermen here are using rods as well as lines, and may be equipped with a spinner or a plug for casting, c. 2000 BC (from P. E. Newsberry, Ben Hasan).

Below: Fish hooks as merchandise. An Egyptian merchant is about to sell or trade his wares. The fact that the hooks seem to be equipped with eyes and have angular shapes does not accurately indicate their age. In Egypt angular shapes appear on wall decorations from the sixth Dynasty, but this picture seems to be of later date, perhaps from the first century BC.

What did the Last Supper consist of? This is a Spanish portrayal of The Lord's Supper, painted on wood in the thirteenth century (Catalonia Museum of Art, Barcelona).

delicacies, and built ruinously expensive cisterns in which to keep noble varieties of fish alive.

Gradually, a clearer distinction began to be made between those who indulged in fishing as a sport and fishing as a livelihood. In a picture of an angler – obviously a wealthy man from Thebes (probably around 1400 BC) a butterfly-like insect has been drawn with symbolic clarity, leaving one to suspect that fly fishing had begun. The Emperors Augustus and Trajan were among the amateur fishermen of Rome.

Fish and dolphins were frequently included in works of art. The Roman baths had a great deal of fish decoration, and, naturally enough, fish also found their way into Church art. The earliest Christians adopted the fish as their symbol. An interesting baptismal font is on view in the church of the St Nilo convent, outside Rome. Here a fisherman with his rod (St Peter?) sits by the Heavenly Gates, fishing up the saved souls in the shape of fish.

In the Nordic countries the 'Iron Age' lasted more than a thousand years, from around 500 BC to AD 600, i.e. to the time of the Vikings.

The transition from wood, shell and bone to bronze,

Above: *An Indian god fishing off the coast of Peru. The picture of the boat of rushes, with its terrifying dragon's head, is a ceramic decoration from the Mochica period (a pre-Incan culture) and depicts the highest deity in combat with the demons of the sea. (v. Hagen,* The Desert Kingdoms of Peru, *London, 1965.)*

Below: *Etruscan fishermen on the sea, detail from Tomba della Caccia e Pesca in Tarquinia, assumed to have been painted around 510 BC. (Reproduced from a drawing in a FAO dissertation by R. Kreuzer,* Fish and its Place in Culture, *1973.)*

iron and steel was not without consequences. The old, basic types of hooks recur, but from now on the shapes become freer, depending partly on the way the metal had been worked. The iron hooks were often bigger than the bronze hooks, a natural development because the boats were stronger and could better be used on the open sea. Norwegian fishermen were venturing far out even in the Stone Age. One of many indications is that bones of ling have been found in the Vistehulene, Jaeren, where the waters are too shallow for ling to be found there naturally. In those days, thousands of years ago, they fished from skin boats. In the Iron Age the boats were bigger and sturdier, built of wood and nailed together.

33

Above: *The fish appears as a symbol in a number of religions. One of the great Hindu divinities is Vishnu, who appears on earth in several incarnations, most often in order to help people. Here we see the incarnation of Vishnu as a fish. The Indian relief depicts the life of Krishna, another of Vishnu's incarnations.* (W. Radcliffe, Fishing from the Earliest Times, London, 1921.)

Right: *The baptismal font in the church of the cloister of St Nilo in the village of Grottaferrata outside Rome. At the time of Christ, baptism took the form of total immersion in water. On the baptismal font in this Greek Catholic church we see people diving into the water, where a metaphorical death awaits them. Those who have been baptized are resurrected in the symbolic shape of a fish. The*

angler sitting with his rod above the Gates of Heaven is, presumably, St Peter. Among the more prosaic details of the font one glimpses above to the left the discarded clothing (the sins) of the person who is to be baptized. Modern anglers will observe that the reel was practically unknown (there are a few exceptions) until recent times. The baptismal font probably dates from the ninth century AD.

One rarely finds old iron hooks that are not damaged. Iron soon rusts under most conditions, and few people save things so apparently insignificant as fish hooks, and then only the archaeologists go to the trouble of studying them. In Norway the oldest finds of iron hooks date from approximately AD 400. One hook was discovered inland at Bykle. Another is from Lista for fishing in salt water. The Bykle hook is especially well preserved. It has a gently curving shank and has been forged with an eye and a barb.

An interesting discovery from a somewhat later period, perhaps one or two centuries before the viking age, has recently been made at Risöya, southwest of Bergen. The finds here reveal that Norwegian farmers so early in history ventured far out to the sea, settling there for seasonal fishing. They must have caught more than they needed for their own use and probably dried the fish since salt was not readily available. Perhaps seafaring merchants travelled about, buying and trading wares. Fish in exchange for fish hooks, for example? The Risöya hooks seem to be made by experts. As far as one can see today all the Risöya hooks had eyes and barbs, the shanks slightly curved. They are of different sizes, the biggest one being suitable for halibut.

The making of fish hooks was gradually left to specialists. The discovery of tools in burial mounds reveals that even before the Vikings much of the finer wrought iron work was done by professional black-

Thor is out fishing for the Serpent of Midgard. With him in the boat is Hyme. Thor is holding his hammer in his right hand, the line in his left. The bait is the head of an ox, its horn serves as a barb. The picture is a relief that has been carved in stone, most likely by Scandinavians who settled in Britain after the Viking period. The relief is to be seen at Cosforth Cross in Cumberland where heathen and Christian concepts are combined.

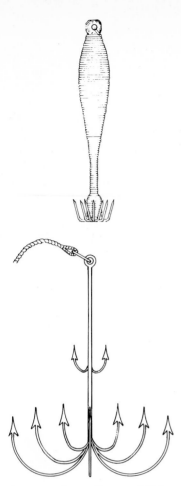

Iron opened up endless possibilities for variations in the shape of a hook. Here are two examples which represent big leaps forward in time.

smiths. There were still many homemade hooks, of course. In fact, in remote areas people have continued to make their own hooks right up to the present time. But, this is rare and, as the centuries passed, the commercial fisherman tended to leave the job of making good hooks to the professional blacksmith. Around the end of the Middle Ages it may be assumed that professional hookmakers were at work far and wide, at least in the coastal centres where fishermen did their buying and selling.

Prior to the invention of printing it is difficult to find accurate pictures of fish hooks from the Middle Ages. But ancient manuscripts do tell us something. Thus, in *The Hours of Catherine of Cleves*, there are handpainted illustrations of fish hooks from the first half of the fifteenth century.

With the development of the printed book, it became apparent that many people of Northern Europe were keen on fishing for leisure and sport. Books on the subject appeared almost simultaneously in The Netherlands, France and England, with Germany following not long after. There are interesting items also in the oldest German book about fishing (and hunting), *Waidmeryk*, which was published in Frankfurt about 1531. The author recommends that the line be made of white (yes, white) horsehair. Perhaps he had been reading Greek or Latin; Plutarch is said to have written the same. The ancients knew better than many fishermen of our time who look down into dark water and think that a dark line is the least visible. They forget that above the fish is a sky just as light as the one we landlubbers see above us. We should remember that white is the protective colour of most fish that expect to be attacked from underneath.

It was in England that the interest in angling first produced a flood of fishing literature. The first thorough treatment of angling was printed in Westminster in 1496 as a part of *The Boke of St Albans*, supposedly written by a woman, Dame Juliana Berners, possibly a prioress. In a part of that book, *The Treatyse of Fyshinge wyth an Angle*, she deals with the art of making hooks. The best hooks are made of needles, she says – the finest darning needles for small fish, embroidery needles for larger fish, and tailor's and shoemaker's needles for

the very big fish. She tells us how to make steel pliable, make a barb, shape and temper the steel (three times heating until red-hot). She recommends that the line be fastened to the inside of the shank, advice that has been repeated by many angling writers.

The accepted publication date of 1496 might be a little misleading. No one knows when the manuscript was originally written. In the Denison Collection there exist manuscript fragments from perhaps the earlier half of the fifteenth century. Between the fragments one finds word for word what was later to be printed in *The Boke of St Albans*.

For a long time afterwards English writers continued to describe the making of hooks. The needles had to be of the best steel from Toledo or Milano. The modern expert will understand the advice given by William Lawson at the beginning of the seventeenth century: 'If the steel is good, the point can never be too sharp', meaning: 'If the steel is not of the best quality the point may break if you make it too sharp'.

In principle the men of the Iron Age were already familiar with the art of making steel from bog iron. But not all iron can be tempered into steel. Steel of fine quality was scarce. Down through the Middle Ages, and long after, the quality of steel was very uneven and good steel was expensive.

Norwegian iron hooks from the Middle Ages. The two on the left were found during excavations in Oslo. The largest of them is 14 cm long, intended for big fish. Hook no. 3 from the left was found during the excavation of the Gokstad viking ship in Vestfold (tenth century). The hook on the far right is one of the poorly preserved hooks from Risöya (seventh century?). All the Risöya hooks seem to have had an eye, not a flat.

Handmade hooks as reproduced in The Treatyse of Fyshinge . . . *from 1496. The hooks are without an eye, a flat (the 'spade') is hammered out in order to keep the leader or the line from sliding off the shank.*

A section of a page from The Hours of Catherine of Cleves, *an illuminated manuscript made in Utrecht in the earlier half of the fifteenth century. The miniature paintings here form a frame around the presentation of Saint Laurentius. He was a guardian saint of the poor, and so the anonymous artist used the motif of fish eating fish: the rich devouring the poor. All the hooks are equipped with barbs.*

The fact that not everyone was in possession of needles of the best quality steel, is borne out by the number of needle cases that women used to carry. Needle cases have been known in Norway for more than one thousand years. They hung from the belt, or were sewn or hooked to the skirt. In them were kept a selection of precious needles, and it was a sign of noble birth of the owner if they were decorated with jewels and silver or some other precious metal.

In the earlier English angling literature there is seldom talk of hook making as a profession. Isaac Walton seems to be one of the first to take up that matter.

The classic book for sports fishermen, Isaac Walton's *The Compleat Angler . . .* , came out in 1653. Most of his wisdom and advice had been gleaned from earlier English literature on the subject. Strangely enough he did little fishing with flies. But Walton writes like a true worshipper of nature and goes through an impressively wide range of the questions a sports fisherman ought to be familiar with. Walton also tells the reader how to braid, knot and prepare a horse's tail, in order to make

Left: *Illustration for the title page of* The Art of Angling *by R. Brookes, London, 1790. By this time the reel had long since been invented, but still in 1790 it was not a disgrace for an expert author to present an angler in an action that is technically so primitive.*

the best possible line. (Threads of silk and vegetable fibre have been twined or spun, but, from time immemorial, sports fishermen have recommended a horse's tail).

When it comes to making hooks, Isaac Walton does an about face. Instead of making hooks oneself, he recommends that one goes to an experienced hook maker. England's best he says, is to be found in London: Charles Kirby in Harp Alley, Shoe Lane, 'the most exact and best Hook-maker this nation affords.'

41

In unknown territory

How many kinds of fish are there? Probably the total number is 22,000 – a figure which two leading experts, Alwyn Wheeler of the Zoological Museum in London, and Daniel M. Cohen, Bureau of Commercial Fisheries, Washington D.C., have arrived at through independent estimates. To be sure, new species may still crop up but, as a rule, only in the fresh waters of Africa, Asia or South America.

If it is borne in mind that one species of fish does not resemble another any more than birds and animals do, that fish choose every conceivable habitat (from great ocean depths to the shallow waters of flooded rice paddies) and, in addition, that the behaviour patterns of fish are highly varied (some fish approach their prey not in order to eat, but out of sheer curiosity, habit or irritation; some strike as they are about to bite, others swallow their victims, still others strain or suck in their food, or use their teeth like a saw or pincers) then it is understandable that anyone who is going to embark upon the production of fish hooks cannot offer only a few types.

In addition, there is one factor which provides a hook manufacturer with an especially hard nut to crack: the fact that fishermen throughout history have been conservative by nature. The fisherman relies on the hook which his father and grandfather has used, or one which he himself has made a good catch with previously. He may try a new method of catching fish, but hardly ever a new type of hook. Age-old traditions have to be broken down before the fisherman will agree to trying hooks other than the ones he is accustomed to

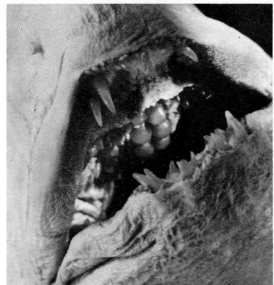

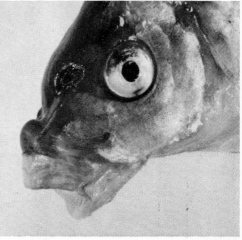

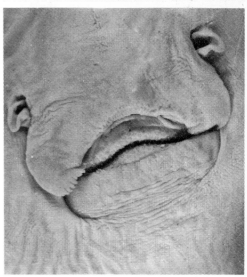

The mouth of a fish, and the way it bites, differs from one species of fish to another.

Above left: *the jaws of the wolffish in which some of the teeth have been wisely transformed into a crushing implement. Since it feeds on snails, mussels, crabs, crayfish, sea urchins and the like, the molars have moved to the centre of its mouth; the six flat knobs (pictured in the roof of the mouth) are teeth intended for crushing.*

Above right: *fish like the bream suck in their food. From its mouth it shoots out a funnel which serves as* a straw. It has no teeth, and so does not eat fish, but creatures that live on the sea bottom. The bream is to be found throughout the northern parts of Europe and Asia. In Norway, bream weighing up to 4 kilograms have been caught.

Bottom left: *The head of a pike. It catches its prey of live fish in an original manner. When the victim is close enough, the pike suddenly darts forward like a torpedo. The victim is attacked from the side and held crosswise in the mouth only to be immediately spat out, caught* again and swallowed whole. From this comes the well-known adage 'Wait until the pike strikes the second time'.

Lower right: *The mouth of the skate is on the pale underside, where the 'face' often resembles that of a human being. However, what looks like eyes and mouth in the picture are, in fact, nostrils. The jaws are flat, the teeth are small and sharp. The skate lives on the bottom, but is an excellent swimmer and an active hunter. (Photos: Per Pethon.)*

43

using, themselves perhaps a heritage from the Stone Age.

There are hundreds of examples, but this one will suffice: in certain parts of the Pacific, the natives have been using a very special type of bonito hook as far back as can be remembered: the shank was of whalebone, with mother-of-pearl to the outer edge. But the point itself was dark and made of tortoise shell. Tortoises began to be scarce and the fishermen discovered a cheaper material instead – plastic. But because the point had always been dark it had to be black plastic. Are we to believe that the bonito in this particular area (Samoa) would not willingly bite if the point were light instead of dark?

But don't think that 'primitive' peoples are the only ones who cling so zealously to tradition. The history of European fishery abounds with tales about conventional thinking and superstitition, and here we may safely include the more 'enlightened' members of the population. From 1721 to 1736, Hans Egede, a Norwegian clergyman, was engaged in an adventurous project as a missionary on Greenland. In his reports from there, he also touches upon the superior fishing methods of the Eskimos. We are told that on the 28 March, 1722, a party of missionaries returned to their camp after a two week journey along the coast. They hadn't managed to pull a single fish out of the sea. They had tried with European fishing equipment, no doubt lines and hooks. But, Hans Egede maintained, the fish weren't accustomed to that kind of equipment. And so the missionaries had to buy the necessary equipment from 'the savages'. Then the fish started biting! These missionaries, who were supposed to represent European civilization, brought equipment which the fish weren't accustomed to and, accordingly, did not allow themselves to be captured by!

All this explains why, in the course of its hundred years, the hook factory at Gjøvik has been operating with such a wide and apparently irrational variety of hooks. It has been calculated that exactly 103,800 different varieties of fish hooks have been manufactured at Gjøvik – each and every one of which has had a singular feature of either material, size, colour or shape which distinguished it from the others. This total

Opposite: 'A Hook Factory' is the title of this picture which appeared in the French Giderot e d'Alemlest encyclopedia in the eighteenth century. Hanging along the walls are tools and coils of wire. The two men who are standing, are busy at the 'benders', which are shown enlarged and in different shapes at the bottom of the page. The barb very likely held the wire in place, when it was drawn round the 'bender'. The job of the third man, who is seated, was probably to cut the barbs or pound out flats. Otherwise, it seems from the picture that the point was cut from only one side.

44

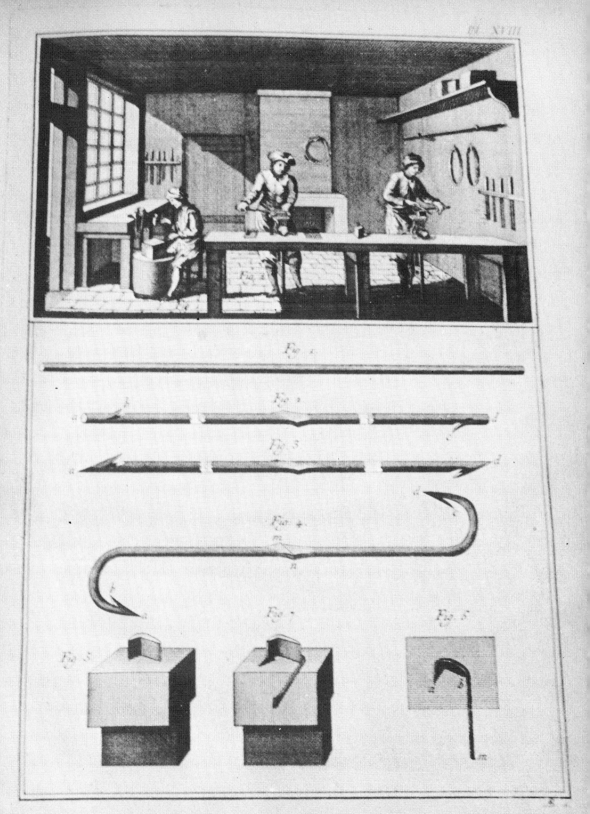

Pl. XVIII

Pêche, Fabrique des Hameçons.

includes all the experiments that have been made. The figure is considerably smaller if one counts only the number of hooks that have actually been on the market. The highest official figure which Mustad has in its advertising is 60,000 varieties of hook. It is no secret, however, that the firm is now making every effort to bring the number of varieties down to a more rational level. One would think that 30,000 would be enough, but in practice it has proved difficult to carry out such a radical reduction.

No one realized what they were letting themselves in for when the production of fish hooks began in 1877. That there were possibilities in the air, was all they knew. Later in the book we shall give an account of the factory and the men who created it. The firm was started in 1832 with the idea of manufacturing steel wire and nails, but soon found itself constantly turning out new products – wire nails and saw blades, tacks, axes, horseshoe nails, and much more besides. The watchword was always the same: replace handwork with machines. In 1877 they got around to fish hooks.

When the production of hooks began, one hundred years ago, there was no technical revolution beyond the mechanical potential of the factory itself. To be sure, one might foresee that good steel would be cheaper in the future: the new French Martin furnaces were epochmaking. But to begin with, it was merely the market and the invention of a hook machine that caused O. Mustad & Søn to take an interest in fish hooks. Little by little, the factory had built up a market for its ironmongery throughout most of southern Norway. This led to a close contact with the coastal towns and the realization that fish hooks were in demand. The skippers from northern Norway especially purchased great quantities of hooks.

The market was obviously big, but also fragmented. Hand-made hooks were to be found throughout the country, in the interior as well as along the coast. Lots of farms and fishing settlements had their own smithies. It is apparent from the hooks, that the blacksmiths followed different procedures. Often the hooks were shaped by bending the iron or steel wire around nails that had been driven into a board. The shape that resulted depended on the position of the nails. If there

The way the fish bites does not always dictate the shape of the hook. Home-made hooks often had angular shapes because it was easy to make them this way. These three hand wrought hooks are from (top to bottom) Norway, China and Peru.

were only a few nails, the hooks usually got sharp corners. In the case of old, hand-forged hooks, one often comes across this shape – in Norway as in other countries.

Traces of the ancient custom of forging hooks by hand remain in many countries, including Norway. This may be due to a purely local preference. In Vågå, it has been a family tradition to use homemade hooks for trout fishing in Lake Gjendin. There the hooks were forged of brass, and were made so thin and pliable that they may be easily extracted from the fish – an advantage when trout go deep and swallow the hook. In such cases it saves time to be able to pull out the hook without having to work it loose carefully.

In the larger cities it soon became apparent that hook-making was a business that could be profitable. Here hook specialists settled down to carry on their trade. In western and northern Norway they called themselves 'angle-makers'. In Eastern Norway they were known as 'needle-makers', which was only natural, since the hook-making profession has often been connected with needle-making.

The production of hooks in Norway must have been quite large. During the first half of the eighteenth century, when Spain and other Mediterranean countries began to purchase dried fish on a large scale, the fisheries in Northern Europe received a powerful stimulus and Norwegian hook makers came into their own. In Alistair Goodlad's book *The Shetland Fishing Saga* (p. 63), we discover that the population of the Shetlands purchased considerable amounts of equipment from Bergen; in 1731, among other things, 8,000 ling hooks and 30,000 haddock hooks. It is quite possible that the Shetland fishermen could just as easily have purchased their hooks in Aberdeen, for example. Perhaps they did their buying in Norway instead, because of various conflicts in the relationship between Shetland and Scotland. It is apparent, at any rate, that the Norwegian production of hooks was well under way.

How early it all began, is not known. In the old 'Citizens' Books' of Bergen, an angle maker is mentioned by name in the seventeenth century. But there were probably several of them, since the earliest trade licenses continually neglected to provide information about profession. In the course of the eighteenth century, the number of hook makers increased considerably. In 1847, Bergen had fifteen angle masters, with seven journeymen and sixteen apprentices. In the course of the decade down to 1863 seven people received their trade licenses as angle makers. Almost all of them had Norwegian names.

But now it was becoming apparent that the time was ripe for a change-over from handicraft to industry. England led the industrial development. The message is clear in this retrospective view that appeared in a Bergen newspaper in April 1918. The headline reads:

Lost Arts
. . . Another is the hook maker profession. A long time has passed since the profession went into a decline, and may now be regarded as having died out completely. But at one time it had considerable influence in this city.

The older citizens of Bergen may remember the hookmaker stalls on various centrally located street corners, and on the Hanseatic Wharves. Different types of hooks were sold, and these were both solid and good, whether the dimensions were large or small.

However, foreign – especially English – factories, through Norwegian wholesale distributors, began to swamp the fishery districts with hooks. And Norwegian manufacturers joined in, so that in a short time, the Masters here in this city had a hard nut to crack in trying to compete. As this could not succeed, the Master Angle-makers had to give up.

The increasing need for fish hooks, around the end of the nineteenth century, was connected with the general upswing in the fishing industry. Fishing harbours resembled forests of masts, the turnover ran into big figures, and there was little time to haggle with individual angle makers over the day's price of hooks.

To be sure, not all the fish were caught with hooks. But if we look at the latter part of the nineteenth century – the time when Mustad's main aim was to enter the Norwegian fish hook market – then hooked implements played a predominant role in the fishing industry. The longline fishermen to the north required great quantities of hooks, and added to this was the use of hooks as a feature of the fishing for food that was carried on daily throughout the country. The fish that were caught with regular hand lines and trolling lines, were

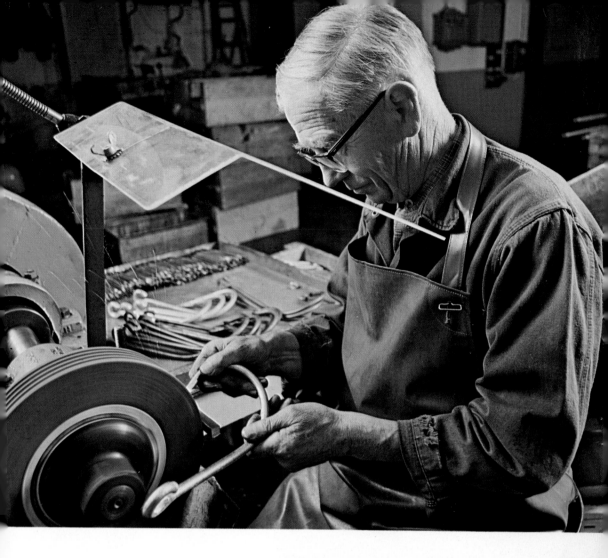

A grinder at work on the expensive shark hooks. The grinding is done before the tempering, and demands painstaking accuracy. As can be seen, the ring of the hook is temporarily half-open, since the hook will later be delivered fitted to a swivel.

Below right: *A glimpse of the tempering process. A tray of fish hooks on its way into the tempering furnace. (Photo: Knudsen Photocentre.)*

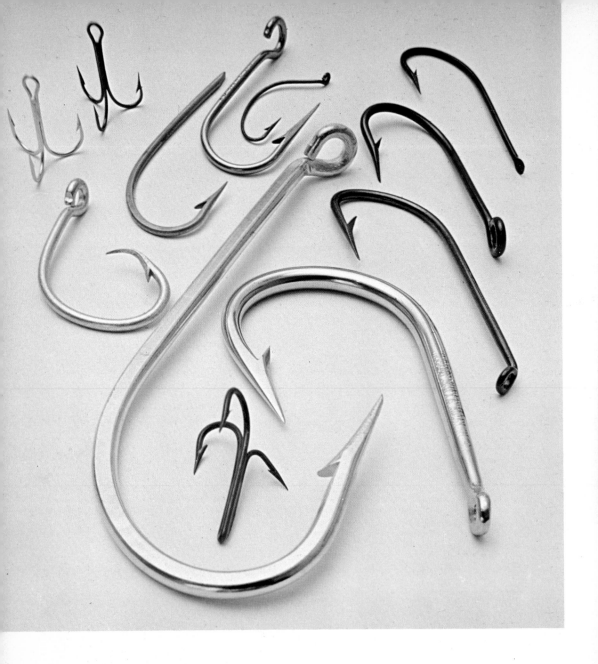

There are Mustad hooks in virtually every colour. Most often the colours are brown, black and tin. But the market demands variations according to local traditions and special purposes – for example, unusual colours intended for spinners, wobblers or shrimp tackles. In many areas, the fishermen insist on blue hooks or even green or red. A considerable number of hooks are nickel-plated. There are also hooks – the so-called 'bright hooks' – that are left in the natural colour of the steel without any finish.

Facing, upper: The hook itself is the least interesting aspect of the fly, but let us not overlook it completely. The three to the right are salmon flies. Three of the flies in the left hand column are dry flies, while the rest are wet flies. There are 76 different Mustad wet flies and 36 varieties of dry flies on the market today. The dry flies require stiffer feathers, usually from the neck of the male – not the female – fowl. Only two of the types are based on reindeer hair and the fur of badger, instead of feathers. In the dry flies, three different types of feathers, as well as woollen thread are used.

Tradition makes complicated demands on the tying and composition of the wet flies. Take a closer look at the salmon fly on the right, a 'Jock Scott'. The part that covers the shank of the hook is called the 'body'. From the bottom up, the body consists of silver twist, ostrich feather, yellow thread, black carded wool yarn, flat silver twist on top. The 'tail' consists of hackles of the pheasant and the red ibis; the 'throat hackle' of the black neck feathers from the male Guinea fowl. The 'wing' is built up of the tail feather of a turkey, and the 'spine' of wing feathers of goose, Florican Bustard, Jungle Cock, King Fisher and Golden Pheasant. (Photo: Frits Solvang)

The Mustad plant at Gjøvik as it appears today. In the background Lake Mjøsa with a corner of the town of Gjøvik far right.

Brusveen Farm can be seen on the edge of the woods in the centre of the picture, the main building painted white. The large rectangular red brick plant on the right houses the hook factory. The smaller, yellow low building in the centre houses the central machine shop.

Top left: *A Roman mosaic in the Tripoli Museum. Surprisingly, we see that the two anglers are using supple rods. Most ancient illustrations depicting fishermen show the rods as stiff poles. The entire scene – not least the landing net – has an amazingly contemporary air about it.*

Bottom left: *In striking contrast to the two anglers of antiquity, we see modern professional fishermen in action. The picture was taken on board a vessel equipped with Mustad's Autoline system. On an Autoline ship, the fishermen can largely work 'indoors', as in a factory. The men here are mending the line after it has come out of the*

splitting machine on the left in the picture. As can be seen, the crew are doing their jobs protected from the weather. Even out on the fishing banks the modern longliners provide a comfort that was hitherto unknown.

Above: *A glimpse of one of Mustad's stockrooms at Gjøvik.*

This sculpture of a fisherman in a
boat was created by Arne Durban,
to decorate the 'courtyard' of the
hook factory. The young lady in
the picture – Kirsten Lysengen – is
the fifth generation of her family to
work in the hook factory and, as
such, was chosen as representative
of the entire staff of Mustad.

Mathias Topp (1840–1930) the inventor whose models of wood and paper paved the way for Mustad's epoch-making first automatic hook machine.

not just for the next day's meat. Many people subsisted throughout the winter on a barrel of salted greyfish or mackerel that had been caught with a hook.

Many Norwegians saw, in the manufacture of hooks, an opportunity of doing business on a large scale. A hook-maker in Bergen (Johannessen), who marked his hooks with a 'J', quoted the size and shape on scales based on 'American models' and 'English models'. From Vardø came the 'Grønbek-hooks', etc.

Down to 1860 it was not unusual to sell hooks without any form of eye or flat. The fishermen themselves hammered out the flat that was needed for fastening the line to the hook.

Mathias Topp was the man who constructed Mustad's first hook machine. This was quite a sensational invention. The wire went in at one end of the machine, and a fish hook came out at the other – for certain marketable types it emerged completely formed.

The fact that it was possible to shape the hook in one

49

and the same machine was a sensation one hundred years ago. Everywhere else in the world fish hooks could only be produced in several stages. In Japan, for a long time to come, the wire was clipped in the factory and the pieces were then sent out to families in the country who earned a part of their livelihood by shaping them. Then the hooks went back to the factory for the finishing touches. Thus, the Japanese hook industry was based partly on handwork. In England, to be sure, machines were used that operated automatically, but the point of the hook was shaped separately.

Even so, the Mustad hooks did not catch on as readily as had been hoped. The English hooks had gained the upper hand to such an extent that the prospects seemed unfavourable for many years. For a long time, the production of hooks at Gjøvik was almost insignificant in comparison with all the other things that were going on in the factory.

The very beginning was exceedingly modest, with only one man for the automatic machine department. Much of the finishing process was carried out manually. During the early years there were no tempering furnaces, so that they were reduced to the old method of tempering the iron in pots of bone meal. The tin-plating was also done by hand. Thus, the old tin pots were not very big. In addition, the first fully-automatic machines could only make hooks with 'spade end' or 'flats'. If the hooks were to be supplied with an eye, then handwork was necessary. Nor did it improve matters that the firm had overtaxed its economy during those first years after 1877. The entire district found itself in a terrible period of depression. For a while things looked bad – and it was necessary to make economies.

But gradually the situation brightened. In 1880 the factory had three hook machines. By 1899 there were eighteen machines that were operated by six men. All in all, there were forty people working in the hook factory. The quality improved and became reliable. This was due, not least, to the fact that from the middle of the 1880's, it was possible to buy steel wire of a new and far better quality. But most important was the experience that had been gained from the realization that a fish hook was a far more complicated product than had originally been imagined.

They soon realised that the hook that comes out of the machine is not fit to fish with. It is so soft that it can be bent by hand, and if it is put into water it will rust before the day is over. It is grey and rough to look at, an eyesore to one who enjoys the sight of the smooth and pretty hooks that are on the market today. It took time to train people to become hook specialists. Another point of difficulty was the tempering process, which had to be reorganised when the new steel was adopted.

It may be assumed that considerable assistance came from English specialists summoned to Gjøvik. The first of these, Henry Haynes, arrived in 1887 – ten years after the start of the hook factory. The next man came only two months later as a specialist in tempering.

It is not easy to know what Hans Mustad, the owner and director of the firm, had in mind at the time. It is certain that Henry Haynes' first job – his official assignment – was to teach Norwegians how to tie flies. The fly factory at Gjøvik also started operating two years later, in 1889. It may be assumed, with equal certainty, that the Englishmen never had anything to do with the hook machines. This invention was closely guarded and it would never have occurred to Hans Mustad to allow a foreigner admittance to the inner sanctum – the automatic machine department.

But, no doubt, he realized that he needed as much English professional assistance as possible. In England, the work was based on centuries of experience in a country with many anglers. This is the way it has always been; the sports fisherman is more interested in quality than in price. For these reasons it was the English hook industry which had acquired the necessary skills.

At the time O. Mustad & Søn was beginning to make a name for itself as a hook manufacturer, British hooks (and particularly the English) totally dominated the world market. By a remarkable quirk of fate, a small village in central England – Redditch – had become the centre of the hook industry. This was true of hooks for sport as well as for professional fishing.

It is likely that considerable numbers of hook makers moved away from London after the catastrophic fire in 1666. What became of Charles Kirby, no one seems to know. Either the plague or the fire might have killed him. One thing is certain however – that the making of hooks was decentralized, a fact that is apparent from the many varieties of hooks that still bear British place names. Leading centres were Dublin and Limerick in Ireland, and Carlisle and Kendal in the English Lake District. In Scotland, the Aberdeen hooks played a significant role. But hooks were also being made on a large scale in Nottingham, London, Harwich, and several other places. It should be added that well-known varieties of hooks may also conceal the names of people: Mr Kirby lived in London, Mr Sproat in Ambleside near Kendal.

For this information we are indebted to an Englishman, Dick Orton of The Angling Foundation, who has had special access to a wealth of historical material. Thus, the starting point is a greatly decentralized hook industry in the British Isles. But why should the unprepossessing inland town of Redditch become the centre of the British hook industry? One explanation relates it to the existence of a large monastery at Redditch. The monks there were reputed to have been skilled artisans – perhaps even metal workers having links with Toledo in Spain. When Henry VIII dissolved the brotherhood the monks were taken in by leading Catholic families in the area.

Originally the whole district was renowned for its metal work – needle making and hook making. The expression 'the whole district' must be emphasized, because originally there was no specific hook centre. In the series *Victoria County History* it is said that the town of Studley in Warwickshire was the centre of the needle industry in the eighteenth century; in those days needles and hooks went hand in hand. Even Birmingham was an influential hook producer, and from there steel wire went out to the needle and hook producers in adjacent towns like Studley, Alcester, Henley-in-Arden and Redditch. It was not until the transition from handwork to industry – around the middle of the nineteenth century – that Redditch became the main hook-producing centre.

The industrial revolution, which can be said to have begun for the needle industry in 1828, proceeded in stages. One invention succeeded another, and even as late as the 1850's, there appears to have been considerable manual work in the hook industry. A great number

Even in industrialized England, there was still the need for considerable handwork in the production of hooks. Here is a 'hook bender', used in Redditch. The factory cut off straight lengths of steel wire in which the barb had already been cut. The bits of wire then went out to home workers who shaped the wire by hand. One sees how the straight piece of wire is pressed onto the hook bender. Even a child could bend the wire into the specific pattern on the bottom of the hook bender.

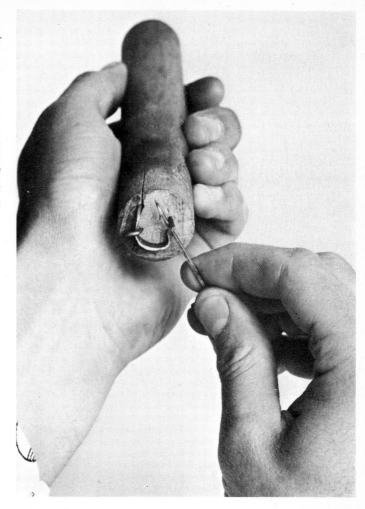

of children were given small implements known as 'hook benders', for use at home. They were so easy to handle that even the hands of a child could bend the steel wire into the shape of a hook.

It was not until the second half of the nineteenth century that the mechanical production of fish hooks was in full swing, and it was then that Redditch gained its reputation. Around the turn of the century, when Mustad salesmen were journeying throughout the world, English hooks were predominant. If one leafs through the reports by the Mustad salesmen in the first decade of our century – whether from Australia, Asia or America – they refer incessantly to competitors, all of whom are reported as being natives of Redditch – Bartleet, Alcock, Milward, Davies, Moore, Sealey,

Hemming, James, Woodfields, Hall, Clarke, Rimmer, Beard and Gould. And there were many left unnamed.

The English workers who went to Gjøvik, had useful information about needles as well as hooks. They were familiar with the art of finishing and also knew very well how a nice hook should look.

A hook machine is like a computer in that both must be programmed. The machine 'stations' have to be tuned in, and only the operator with an experienced eye is able to ensure that a proper and exact shape will come out of the machine.

Henry Haynes settled down in Gjøvik for good. He came with his wife, son, and six daughters. As far as is known, all the daughters married in Norway, one of them to a son of Mathias Topp. A tiny English colony grew up in Gjøvik when, after two years, Henry Haynes went to England and brought back no less than nine skilled workers. Two of these were women, both fly tiers. These English craftsmen (who later returned home to England) were to assist at the start of Mustad's needle factory. They were all specialists in different trades – tempering, scouring, grinding, polishing – and the hook factory needed this professional knowledge just as much as the needle factory did.

The hooks had been given English names right from the start. Among the first series of hooks to be produced were the Harwich hooks, which are still marketed today. Since the firm concentrated on the Norwegian market to begin with, Norwegian geographical names were sometimes used. In the Kristiansund Museum, for example, there is an old Mustad showcase, of the kind that was sent out to the larger stores that sold fishing equipment. In this showcase, one variety of hook is called 'Harwich, Aalesund Model'.

With English workers at the factory, it is no wonder that a number of English expressions drifted into the daily language among the Mustad employees. Traces of these English words are still to be heard in the factory today. Like our seamen the hook makers at Gjøvik seem to prefer English words to Norwegian ones. These English expressions, mostly technical, have been transformed into Norwegian dialect through several generations and would hardly be recognized today as English by an Englishman.

The most stubborn resistance came from the longline fishermen in the north of Norway. This line boat appears to be modest, but each vessel used large quantities of hooks. As the man in the bow is pulling in the longline, his companion takes the cod off the hook. Some fishermen cut the leader and allowed the hook to remain in the fish's mouth. Under certain conditions, they even cut off the head and let both hook and head fall in the sea.

The company soon found how difficult it was to force their way into the domestic market.

The fishermen examined the Mustad hook with a critical eye and discovered certain things that were unfamiliar to them. On most of Mustad's machine-made hooks, the point itself was flat and not round as it was on English hooks and on most hooks that had been ground.

It did not help that Mustad put English labels on the packages of hooks. The greatest resistance came from the big customers – the longline fishermen in the north of Norway. So doggedly did they refuse to accept the typical shape of the Mustad hooks, that the factory, with considerable difficulty, had to construct special machinery just so the point would resemble a rounded point. (Mathias Topp visited the Chicago Exposition in 1893, and it was there that he was supposed to have got the idea for a solution to the problem.) Ironically, when the market had finally been conquered, it no longer made any difference what the point looked like as long as it was sharp enough.

The decisive upswing occurred in the period between

the wars. Then the big orders began to pour in, and the trade acquired completely new dimensions. In the 1920's it seemed obvious that Mustad hooks were well on the way to achieving a position as one of the leading brands.

How did this come about? Later in the book we shall touch on the quality and reliability of the hooks. But one special aspect will be examined here. In the field of marketing, O. Mustad & Søn was a pioneer. When the directors realized that the hooks really were capable of competing, they established international goals and invested in the future with an astonishing degree of daring and imagination.

Of course, their existing trade proved a sound basis for expansion. The horseshoe nail in particular had become a world article that had provided the Mustad factory with an exceptionally good insight into market conditions abroad. The need for horseshoe nails was enormous before the automobile caught on in earnest. This had resulted in sales in widely-scattered areas from the Balkans and Turkey to South America and Australia. The horseshoe nail was the most important commodity, but it also helped the sale of hooks when Mustad salesmen scoured the world to find buyers for the firm's products. During the first decades of the twentieth century, the salesmen not only took samples of nails with them, but also other commodities such as spikes and fish hooks. Not until the inter-war years did the firm change its tactics. Then, the salesmen of horseshoe nails and those of fish hooks were to part company.

But other factories must also have had the same initial advantages. A considerable number of the English hook factories made more than fish hooks – for example, needles of a quality that were and are in demand throughout the world.

No doubt Mustad developed a unique sales technique. Right from the beginning the salesmen never hesitated to send home samples of the types of hooks that were preferred locally in every district. It appears that the salesmen followed a certain pattern, they took care not to recommend their own products in general, the most important thing being to discover which types of hooks were marketable in each place.

Another characteristic was the way the Mustad salesmen travelled. They set out into unknown territory, hunted up the fishing population, and stayed far away from the larger trading centres. Their travel routes stretched like spider webs across the globe. Here is an extract from a salesman's letter home to Mustad from Lima, Peru, in February 1911:

Because of military duty, I have to be back home no later than July of this year, and as I must count on a month in Japan, the route hereafter will be somewhat different. However, I think I will be able to cover parts of Central America, Colombia and Venezuela. According to Mr Halfdan Mustad, I am not to waste time on Mexico now. Please send me interesting assignments without delay to San Francisco.

From home, the same salesman received his instructions:

We understand that you will be coming home by way of Japan, and as you will apparently be returning via Vladivostok, we would like you to visit some firms there. Considerable quantities of hooks go to Siberia, especially via Moscow firms, which partly have branches over there, but hooks are sold through Hamburg exporters as well.

In a letter from Buenos Aires, a salesman describes the way he will be travelling through South America in 1912:

I plan to arrange the following route: From Antofagasta to Bolivia (Uguni) by train; from there by mule to Potosi and Sucre. From Sucre I will probably take a mule to Cochabamba, from there by mule to Oruno, then on by train via La Paz to Mollendo. I will be arriving in the middle of the rainy season, so . . .

Many of the salesmen's reports back to Norway may be read today as interesting historical documents. The following is a report from Japan in 1911:

The Japanese are very suspicious vis à vis foreigners. They wish to get them out of the country, the few that are left.
If you try to get information of any kind, they will not understand. A great difficulty here is the language. There are very few Japanese who speak or understand English. They are proud and selfish, demanding that you shall learn their own language. I send you from Yokohama a complete

collection of six types and sizes that are most commonly used. The Japanese fishermen, however, find the Japanese hooks too soft. As you will see the spades of the hooks have a very special form. There is no numbering of the Japanese hooks, only the length of the hook is stated by Japanese measure. The Japanese labels are without any firm name and show only the price and what kind of fish the hook is meant for. In Osaka I found over 100 different types of hooks, but I am told that there are on sale in Japan altogether 3,000 hooks of different type and size.

Concerning Rio de Janeiro one learns that this city (1911) has

over 800,000 inhabitants and has undergone a tremendous expansion during the last 3–4 years. The many horsedrawn trams of a few years back have now vanished and have been replaced by electric trams. Delivery automobiles are used on a large scale, and now there are also special rail lines for electric vans.

A salesman's report from Bombay in 1909, states that the approximately one million inhabitants are living in the most indescribable squalor. 'No one understands English, not even the police. The most well known hooks here are Woodfield's and Milwards'.' Concerning Calcutta, he writes that the potential market for hooks here is very big, so that every order from here must be filled out 'with the greatest of care'. The salesman plans to visit a large number of villages along the rivers that are abounding in fish. 'As a matter of fact', he adds,

there is such a multitude of crocodiles in these rivers that fishing from small boats is not without danger. Europeans have bigger and better boats, however, and they shoot the crocodiles for sport at the same time as they fish. Both German and English hooks are sold here, especially Milward. A considerable number of hooks carry an iron hand as a trademark, or Cross Fish brand. Some are marked with a H on the plate, very likely from Hemming & Son. . . .

From each new place the Mustad salesmen sent home samples of the hooks that the fishermen insisted on using. Then it became the task of the Gjøvik factory to analyse the samples, in order to be able to send the correct type of hook to the correct place the next time.

Going through the salesmen's reports from the first

decade of this century one is surprised by the small variety of hooks that were sold nearly everywhere. In America as in Australia and Asia the hooks offered for sale were, as a rule, either straight, roundbent or Kirby Sea – 'Best Steel'. Japan must have been an exception, and here they also preferred more shiny hooks to the tin-coloured ones. Thus, Mustad got an advantage by always looking at the local fishermen's traditional and very special likings.

The biggest difficulty for the Mustad salesmen seems to have been the many trademarks that were on the market. Often the salesmen were told that it would be a useless attempt to introduce a 'Key brand' hook, because the local fishermen would stick to those trade marks he was acquainted with – for instance, trademarks with a lion, a star, a swan from Redditch.

Price and quality were, of course, important factors. German hooks found their way in great quantities to South America as well as to Africa and Asia. They were extremely cheap, but of bad quality. On the other hand fine British hooks were often sold at high prices, sometimes 'absurdly high', an expression used for instance of the Hemming hooks sold in Havana (where United States hooks made by A. Johnson, Brooklyn also competed).

It appears as if the Mustad salesmen had an unusually free hand. One explanation is, no doubt, the fact that the factory's directors had themselves been salesmen, and thus understood what a free hand could mean. Before the First World War, for example, Hans Mustad's youngest son, Christian, could be found sending sales reports from North America, Egypt, Poland, Athens and Moscow. With every new trip, the directors insisted that the next trip had to be even longer, and the traveller's instructions had to be expanded. As the years went by, O. Mustad & Søn sent out people – not merely to get orders for hooks or to handpick agents. Many were sent out for the sole purpose of informing the firm about possible markets in the future.

A striking example is a journey to Africa in 1929. By and large, Africa was a part of the world that was relatively unexplored. The salesman who had been assigned the job, spent no less than 445 days crossing the continent with no other important task than to inform the directors about the potential sale of hooks.

From Dakar, he followed the western coast down to the southern tip of Africa. From there he headed north again, along the eastern coast through Somalia and Ethiopia, all the way up to Egypt. Along the way he made laborious detours up rivers like the Congo, Niger, Zambesi and Lualuba, to the great inland lakes of Africa.

Only ten years later, large parts of Africa had been transformed; modern communications had appeared, hotels with European comforts had been built, and a public health service had been established which removed much of the risk of contracting the deadly tropical diseases. In 1929, conditions were still so primitive that the traveller was often made to feel like a Livingstone on a journey of discovery.

Along the west coast, where a travelling salesman in those days was reduced to travelling by freighter, it turned out that the market had been amply covered by Japanese, German, French and British hooks. There were 'Portuguese' hooks as well, but these had probably been made in England.

Few, if any, harbours had quays. Anyone wishing to go ashore, had to be carried through the breakers in native surf boats – a landing that was not without danger when the traveller had baggage weighing 100 kgs. or more.

The journey into the interior was by slow-moving flat-bottomed riverboats. Nights were spent on the river bank, largely because the currents could be rough in African rivers and often carried along islands of trees and vines which could be dangerous in the darkness.

The riverboat up the Congo took twenty-one days to get to Stanleyville. Here is an excerpt from the salesman's notes:

It takes a terribly long time. Every afternoon at four we tie up for the night, as a rule by a village, which is nothing more than a clearing in the jungle and has 40–50 inhabitants. Here, on the equator, it becomes as black as pitch in 10–15 minutes. The natives in such villages are quite wild, and seldom know what a fish hook is for.

There are no shops here. Everywhere I have been, I have handed out a few bags of samples among the natives, I have tried in vain to get hold of hooks which have been made by

the natives. The only thing I have come across – a kind of harpoon – is being sent separately – but it is hardly a commercial article.

In some places I have seen the natives using small casting nets. The most uncivilized use neither hooks nor nets, but make fish traps of bamboo fibre, or build a kind of trap out of bamboo by the water's edge. The natives wade out and catch the fish with their hands.

The Congo River is abounding in fish, it is said that 600 different varieties are to be found here. The largest one I saw that had been caught, was a metre and a half long, and actually resembled a cod. But the taste was rather insipid.

The journey up the river is quite monotonous. Always the same thing to look at, impenetrable jungle down to the water's edge. The only variations now and then, are the tiny villages that appear from time to time, as well as some enormous crocodiles that lie sunning themselves on sandbanks in the middle of the river. It's fun to see them dive as quick as a flash into the water as soon as they hear the noise of the riverboat.

Less entertaining are the mosquitos that appear in great clouds as soon as it grows dark. The tsetse fly is also very annoying. There are hundreds of cases of sleeping sickness among the natives, and in Kinshasa, seventeen whites are down with the same sickness after having been bitten by the tsetse fly. In large places like Coquilatville and Stanleyville, one has to obtain a 'certificat medical', and if, as a European, I do not go to the doctor, I risk a fine of 2,500 francs.

To the best of my ability, I have advertised Mustad's fish hooks and handed out samples that have turned out to be most welcome – according to the circumstances. I especially remember what happened in a little village, the name of which – if it had one – certainly does not appear on any map. When the boat had come close to the shore, a group of adults and children gathered expectantly, and wondered what this white man was up to who stood by the rail rummaging in a bag. Then it happened. I threw some 20–30 fish hooks down on the shore. A veritable fist fight broke out over the hooks. Here, I thought, they must be accustomed to using fish hooks, and fish hooks must be scarce. Great was my astonishment, however, when the fight was over. Everyone who had got hold of a fish hook scattered. But what did they do? Were they going to fish? No, they sat down, and with the tip of the hook they picked out sand worms that were half a centimetre long from under their nails. All the natives go barefoot, and sand worms are supposed to be quite bothersome.

'I have seen the natives catch fish that weigh up to 60–70 kgs . . .' The African fish here is of the perch family, caught in Sudan in the white Nile (FAO photo). River fish can be much bigger than this. In Russia a beluga was caught that was 7·5 metres long and weighed nearly a ton and a half. The beluga is the world's largest bony fish, a kind of sturgeon that spawns in rivers. Among the largest fresh water fish is also the European catfish, which is still found in Sweden. In 1871 a European catfish on the fish market in Eskiltuna was 3·6 metres long.

On the 31st of March, 1930, we arrived in Stanleyville. There was one hotel there, the Hotel des Chûtes. It had neither electric lights nor running water. A large number of fishermen have installed themselves by the Stanley Falls, not far from the town. They do not use hooks, but build large, funnel-shaped traps (des nasses) of bamboo. The opening is up to two metres in diameter. In quite an ingenious fashion, the natives have built up a scaffolding of sorts. The fish that come too close to the falls are caught by the great masses of water and carried straight into the traps. I have seen natives catch fish weighing 60–70 kilograms in this manner. The method is not without danger to the natives. They balance on the primitive scaffolding 15–20 metres over the waterfall, and it happens, not seldom, that they fall out into the falls and are lost.

Even in those days – around 1930 – it would not do for a white man to travel without a 'boy'. It was said that 'the white man would lose all respect if he carried his baggage himself'. On this journey, the Mustad salesmen changed 'boys' thirty-two times. There was constant tribal hostility, and the boys refused to accompany him out of their own territory. These native 'boys' were helpful, good-natured and kind – even though fate did not always thank them for it. One of the boys was eaten by a crocodile. He was busy washing his master's shirt. When the white shirt fell in the river, the boy dived in, without hesitating.

The experience this salesman gained of fish hooks in Africa is testimony to the theory that Africa is one part of the world in which the use of hooks has not been widespread. The first person to record this fact was David Livingstone. He found it strange that a single fish could make its way up the Congo river, full as it was of weirs and nets and all kinds of traps.

But Africa is big. And if it may generally be said that Africa has never been 'hook country', and that the use of hooks was largely due to European influence (and also, in part, Japanese and Arab), it is a rule with many exceptions. From ancient times longline fishing has been carried on in the Niger River. The lines were stretched along as well as across the river, in both instances with special hooks that were not baited, but in which the fish became entangled during the night, as if they were nets. In his book *Tropical Fisheries* (Camelot Press, London, 1961), C. F. Hickling maintains that modern longline

The fishing at Stanley Falls is part of a long tradition which has continued down to the present. This photo was taken after World War II. The whole arrangement seems dangerous for the fishermen – just as described by the Mustad salesman in 1930.

fishing may have had its origins in Africa. In the large lakes of Tanganyika and Victoria, longline fishing is supposed to be especially ancient.

Africa, in fact, became a considerable market for Mustad hooks. An order for hooks, which is among the strangest in the 100-year history of the firm, bears witness to the fact that the firm was not unknown on this continent: it happened during the years after the Second World War, when the great Kariba Dam was under construction in the borderland between Gambia and Rhodesia. As the work progressed, an unforeseen problem arose with the rising of the water. It turned out that a number of rhinoceroses had been trapped on islands. As the rhinoceros cannot swim, it was necessary to find a way to move the animals to the mainland. And so the hook factory at Gjøvik received an unusual order. Could the factory, as quickly as possible, make a considerable number of hypodermic needles with barbs?

The order was fulfilled and the rhinoceroses were duly shot at with the anaesthetic needles and the animals were saved. The sedative worked as soon as they had been hit; the animals could stand on their own feet, but behaved as if they were asleep.

A sequel to this story is that the rhinos turned out to be quite ungrateful when they woke up again in unknown territory. They tore down trees and were said to have gone on a rampage.

The story of the natives who used fish hooks to clean their toe nails – apparently because they knew of no other use for fish hooks – has several parallels. In South America, it happened in the old days that Mustad hooks were used as jewellery. Women who were fond of dressing up, thought the nice, shiny fish hooks were pretty and used them as ear rings. The barb kept them in place.

It was sometimes difficult to persuade people that fish-hooks were a commmercial commodity. This is what a Mustad salesman experienced in South America:

Between March, 1927 and October, 1928, I was on a journey in Latin America. One day I came from Cuzco, in Peru, to the border town of Puno in order to take the boat across Lake Titicaca to Bolivia. But in Puno I was arrested. The police found sample cards of fish hooks in my baggage and my explanation that I was travelling for Mustad who manu-factured fish hooks and other articles, was not accepted. One did not travel such a long distance just to sell fish hooks, so something had to be wrong as far as I – and my occupation – was concerned. I was detained for three days. I was allowed to live at the hotel, but had to report to the police station morning and night.

As far as one can gather there was never any penny-pinching over the travel expenses. One salesman, who was later in charge of exports, estimated that he spent a total of fourteen years of his life travelling. He visited Canada and the United States twenty-six times, and went regularly to the Far East, including India, Pakistan, Indonesia, Burma, Vietnam and Japan. In Indochina – now Vietnam – he discovered, to his surprise, that hooks of steel did not exist. They used home-made hooks that were made of brass or bronze, and, accordingly, were easy to bend. The Vietnamese were greatly impressed when they were allowed to test the strength of a Mustad steel hook. This same salesman received his biggest order in Calcutta – a total of 60 million hooks in a single shipment.

There is every reason to praise the vision of those in charge of marketing, but recognition should also be made of the firm's numerous travelling salesmen. They faced toil, adversity and hardships to a degree that is unparalleled in our modern age. Many of the journeys were like minor expeditions. The baggage contained not only tropical clothing with high mosquito boots, but also formal dinner dress. A salesman in equatorial regions had to remain on his guard and be careful how much he drank. It was dangerous to go to bed even when one was exhausted without examining the mosquito netting.

Travellers often took their beds with them on a journey, to be hung up on hooks on the walls. The only things of comfort the small lodging houses and state bungalows had to offer were a floor and a zinc basin to wash in. The concept of comfort has changed considerably in the course of the past fifty years. Those were the days when Mustad's travelling salesman in New Zealand could find hotels in which the following notice advertised how excellent they were:

BED 1/-   CLEAN BED 1/6

Hooks for the world market

If the reader is good at arithmetic, and knows that seven people can sit around a table in 5,040 different ways, then it is easier to understand why Mustad has found it necessary to make such an enormous variety of fish hooks. In every single family of hooks (and there are plenty of families) there is a multitude of possible variations.

Let us start with the variations in colour. The hooks that are most often seen, are tin-coloured, black or brown. But take a look at Plate II on the colour pages. All the hooks are made by Mustad and cover virtually the entire colour spectrum – from blue, red and green to brown, black, nickel, tin, gold, copper and 'bright'.

The bright hooks are unfamiliar to many people. They are delivered without finish, polished in the colour of steel. They rust quickly, but there is a market for them even today. Strangely enough, in certain areas in America there has been at the same time a market for bright hooks and for the far more expensive nickel hooks, for the same fish (salmon) and the same method of fishing (line). These most expensive hooks, made of M-Nikkel wire, are guaranteed 'rust proof under absolutely all conditions'. A manufacturer of pork rind bait equipped certain categories of his baits with a tail hook. Since pork rind is packed in glass jars filled with salt brine, the M-Nikkel was found to be the only hook material that would not corrode.

Before discussing further aspects of colour let us have a look at the hooks themselves. Some examples will illustrate the incredible number of varieties that a hook manufacturer has to make.

As is apparent from the following pages, there are thousands of different types of fish hooks. And there are other variations that have not been touched upon.

Quite often one hears fishermen claim that they like this or that type of fish hook, but wouldn't it be possible to make this particular hook a little thinner, or with a slightly longer shank? As a result it becomes necessary to produce one series after the other with wire of different thicknesses and with shanks of various lengths. There are specific designations for such deviations. If, for example, it says on the label: 2 × Short Shank, this means that the hooks are three 'numbers' shorter than standard length (the order is Short Shank – × Short Shank – 2 × Short Shank).

The same is true of the thickness of the metal wire. The scale can go from 7 × extra thick wire down to 5 × for the thinner gauges. It may seem strange that everything under the sun should be supplied according to the whim of the fishermen. But the customer is always right! The boxes of 7 × hooks go almost entirely to India where there is a strong river fish, the Mahseer, which is much coveted by anglers. It is caught with a small spoon bait – which has to be equipped with extra strong hooks; and so one is asked for a '7 Extra Strong'.

To complicate matters even more, some fishermen prefer the hooks to be delivered with the eye half open, and pliable so that the fisherman can close it himself. In some places the fishermen want the hooks to be especially soft. This is true of Nova Scotia, for example, where Canadian line fishermen have their own technique. In order to save time taking the cod off the hook, they want hooks with such a pliable bend that they can wrench off the fish. As they jerk the hook straightens out. But surely they ruin the hook? No, the fishermen stick the hook into a small tool on the gunwale, and simply bend the hook back to normal shape. The hooks are otherwise standard enough – typically a 6/0 Limerick.

If you have ever thought that a fish hook is just a fish hook, then think again. Shown on this page are the most typical variants of the points and the shape of the wire.

Top right: *Five of the most common shapes of the points.*

The two main types are 'Hollow point' and 'Superior point'. The latter is identical with 'Spear point'. Seen in profile, the outer edge of the Hollow point is straight, while the line curves between the point and the innermost part of the barb. A typical Superior point is exactly the opposite: While the outer edge is curved, the line is straight from the point to the inside of the barb. A 'Dublin point' has features of both main types. A 'Curved in point' means that the point is curved in towards the shank.

To the right we see how the bend can be either straight or offset. On a 'kirbed' hook, the offset is to the left if you hold the hook with the bend up and the point facing you. On a 'reversed' hook, the offset is in the opposite direction.

Bottom right: *a cross section of three different shapes of wire. A regular hook is made of round steel wire. In a 'forged' hook the wire is slightly flattened instead of round, so that the hook looks wide from the side and narrow from the front.*

Extra strong hooks are usually 'forged'. A 'knife-edge' point is best illustrated by a cross section; the wire is shaped so that it tapers like the blade of a knife from wide to narrow.

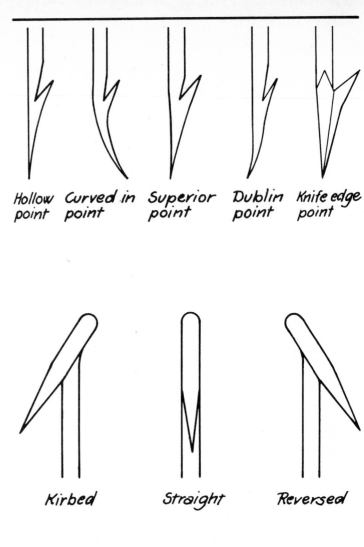

Hollow point   Curved in point   Superior point   Dublin point   Knife edge point

Kirbed   Straight   Reversed

Regular   Forged   Knife edge

Top right: *Seven types of rings and flats.*

(1) A 'ball' ring.

(2) On a 'tapered' ring, the thickness of the wire is reduced. It gradually tapers towards the end of the ring.

(3) Hooks with a 'looped eye' are used, among other things, for tying salmon flies.

(4) Hooks with a 'needle eye' are used especially by big game fishermen when it is essential that the bait fish is damaged as little as possible when pushing the hook through it.

(5) A 'marked shank' is reminiscent of several bone hooks of the Stone Age, which had notches along the shank to facilitate fastening the hook to the leader; with the introduction of nylon, marked shanks have become less popular.

(6) & (7) A 'flatted shank' has a flat or spade end instead of a ring. Flats were regularly used in commercial fishing, until nylon made rings more practical. But the flat is still in general use. Sometimes the fishermen insist on a flat with a hole in it, especially for long line tuna fishing.

Right: *Examples of the different ways in which the ring may be placed in relation to the shank. From experience, most 'turned up' hooks are used for tying dry flies, and 'turned down' hooks for wet flies, which does not prevent individual anglers from swearing by the 'turned up' hook for wet flies as well, inclusive of streamers. 'Sliced hooks' have from one to four barbs on the shank, which helps to keep the bait in place, for example roe, worms and mussels. Hooks with a 'tip shank bent back' are intended primarily for fishing with worms; this type goes mainly to the USA, especially Ohio.*

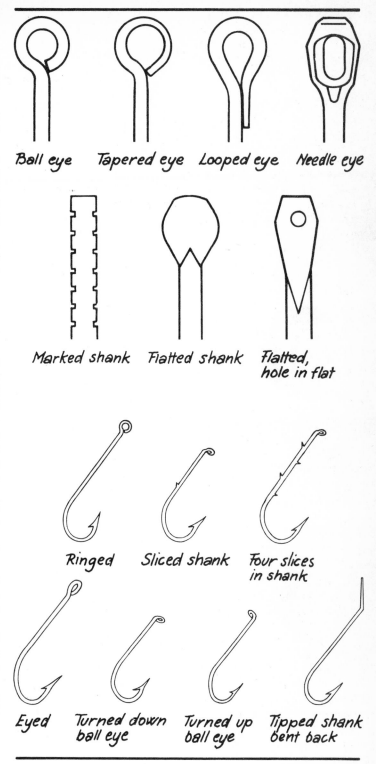

Ball eye   Tapered eye   Looped eye   Needle eye

Marked shank   Flatted shank   Flatted, hole in flat

Ringed   Sliced shank   Four slices in shank

Eyed   Turned down ball eye   Turned up ball eye   Tipped shank bent back

How different fish hooks can be

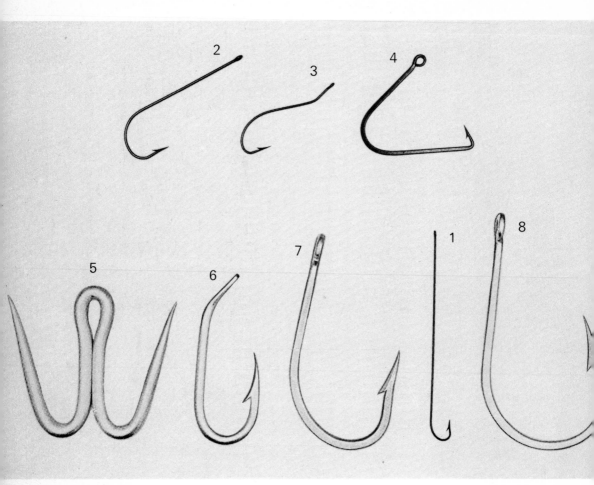

The Mustad hooks that are shown here have strikingly different shapes. Obviously the hooks here are not shown in true sizes; some of these hooks could be for fishing for trout, others for shark.

(1) A Bridgeport Snapper, qual. 3376. It has a distinctively narrow gap and a long shank. A Chestertown hook could be an equally serviceable example. Both are mostly used for catching flounder, particularly in America.

(2) An Aberdeen hook, qual.

3261. This hook with a wide gap is especially in demand in Sweden and the USA.

(3) A Central Draught, qual. 3777, used largely in America. Here the shank is broken. The shank can be broken in different ways, or equipped with slices, depending on the fisherman's taste and to what extent he wants to make a 'weedless'.

(4) This is a pike hook, qual. 7724. As a rule it is baited with small fish.

(5) A hook, qual. 22560, for capturing alligators and crocodiles. The reduction in size here is obvious; the wire of this hook is $\frac{3}{8}$ of an inch.

(6) This hook, qual. 9202, with round wire and a long flat with a hole in it, is used extensively for commercial long-line tuna fishing.

(7 and 8) Both these hooks are typical big game hooks.

(9) The name is Round Haddock, qual. 2362, but the hook is not typical of those

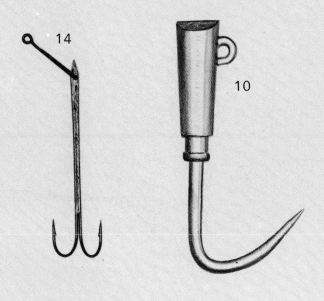

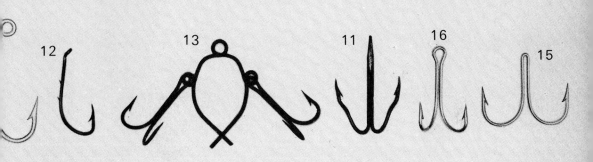

used for haddock fishing in northern waters. It is largely in demand in the Philippines; the hook is eyed.

(10) Qual. 9402 is for tuna, especially bonito. Covered with feathers it is used for trolling. The hook has no barb and is not baited.

(11) This Mustad Double, qual. 9416, the so-called 'parrot hook', has its main market in North America.

(12) The shank of this hook, qual.

92641, is sliced, but characterized by the curved-in beak point.

(13) This 'Live Bait Steel Mounts', qual. 7149, is an example of a 'standing' or 'steady' hook, i.e. an implement that can hang without supervision, either fixed on land or to the ice.

(14) On this Double Live Bait hook, qual. 9418, the shank is tapered, gradually becoming narrower and has neither a flat nor a ring. A 'tapered'

shank is normally meant for hooks that are to have a special 'snelling'.

(15) This double hook, qual. 674 C is often used as a component in moulded gigs, etc.

(16) Here is another double hook, a so-called 'loose double', meaning that the shanks are not soldered together. This particular version, qual. 7826, is nickel-plated and used extensively in North America.

How similar fish hooks can be

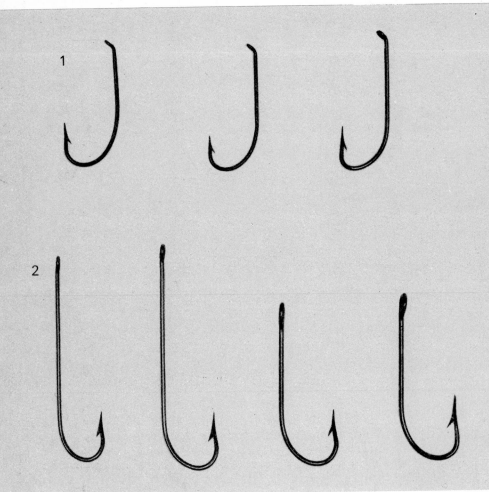

Without closer examination, it may sometimes be difficult – even for an expert – to distinguish between two hooks that have different family names. Here are several examples, presented in groups, with the names listed from left to right.

(1) The arrangement with quality numbers may sometimes lead to the result that a Sproat qual. 3904 A, for example, will resemble a Limerick qual. 9578 A. Both have comparatively short points and short barbs.

The Sproat is one of the most difficult hooks to shape, the bend merging imperceptibly into the shank. It is also characteristic of the Limerick that the bend extends a good distance up the shank, but on a Limerick the sharpest curve is at the front end of the bend. In this instance, one should perhaps refer to a Sproat that is adapted to the Limerick shape. The third hook is a typical Limerick.

(2) Shows a Kendal Kirby, qual.

4540½, compared with a Carlisle qual. 3191. The bend on the latter is somewhat more sharply-defined than on the Kendal Kirby hook, but the difference in shape is insignificant. Far right: two hooks for comparison. First, a Cincinati Bass Hook, qual. 3305, and a Best Kirby qual. 3136. The difference is so slight that specialists sometimes require a micrometer to tell them apart.

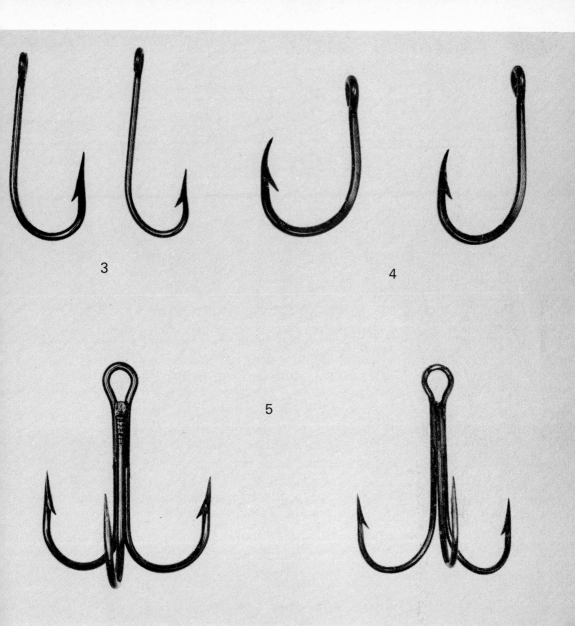

3

4

5

(3) Side by side: *A Superior Bowed, qual. 1608, and a Superior Best Kirby, qual. 802. Only a trained eye will see at once that the Superior Bowed hook has a slightly wider gap.*

(4) *The hook on the left is a Mustad Beak hook, qual. 92081. The other hook is* known by its quality number, 530 A, and is typical for the French market. It was formerly known as a 'Hameçon Rond-Mustad'. In order to be able to identify a hook, the specialist often has to measure such a thing as the thickness of the wire. In this case, one would find that the wire of 530 *A is slightly thicker than that of qual. 9281.*

(5) *It is often difficult to distinguish between two treble hooks. Left: a treble qual. 35647, Right: 36153. The shape of the first hook has a perfectly round bend and a wide gap. (Photo: Frits Solvang)*

The same type of hook in different sizes

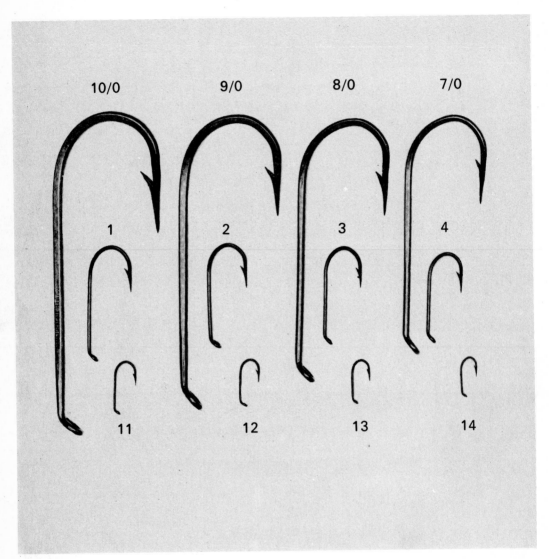

On the preceding pages, we have shown some possible variations in the shape of a fish hook. In addition to this, there is the variation in size. Practically every hook is produced in a series in which the size is indicated by a specific number.

There is no record of the reason-ing behind the numbering system; why large hooks, as a rule, are designated by a zero behind the number, while the smaller hooks are generally given numbers with-out the zero.

Unfortunately, there is no ab-solute measurement that can pro-vide the exact size of every single fish hook, but we shall return to this question later in this chapter.

Let us first see how a whole series of hooks (the Viking Series, qual. 7958) can be numbered. Each hook in the series is reproduced in its actual size.

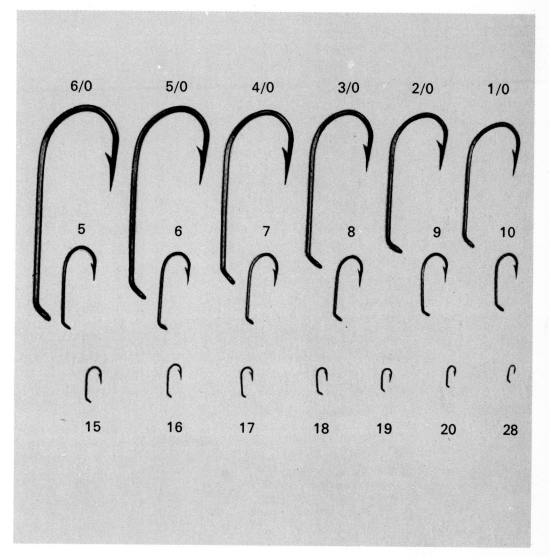

The Viking Series is used in practically every part of the world. Like the series qual. 9480 these hooks are largely used for fly tying.

This is a tapered eye hook. The bend is 'reversed', but this does not necessarily mean that a hook with an offset bend – 'kirbed' or 're-versed' is preferable. The question of which is the best, a straight or an offset hook, is highly debatable. The flies that Mustad tie them-selves, are generally tied on straight hooks. The smaller hooks in this series are not made for curiosity's sake, as one might imagine. All sizes above are sold and actually used by anglers. The maximum length of the tiniest hook, number 28, is 4·5 millimetres. (Photo: Frits Solvang)

THE COLOURS

Most of the hooks are either light or dark. Some people want hooks that are as shiny as possible, while others choose dark colours. The basis for preference comes from the answer to the question: Should the hook be visible or not?

The answer is a qualified one: it depends on the circumstances. In general, the fisherman believes that it is the bait that should be seen, not the hook. This is usually the thinking of the line fisherman. It is also obvious that a fly ought to resemble a fly and not a hook. But, with equal certainty, it is known that something shiny can attract a fish. During trolling, the spring mackerel will take an empty hook with a voracious appetite. The hook need not be shiny; it is enough that it moves in the water.

If the starting point is that the hook should be invisible, then there should be as many colours for hooks as there are colours in nature, in the water, at the bottom, in the trees, in the sky.

Cynics say that gold catches the fisherman, not the fish. But does this hold true today? There are better finishes against rust than gold, because iron and gold are not the best of friends, chemically speaking. But it is a fact that hooks with a coating of pure gold fish well, and there has been a greater demand for them in recent years. A considerable number of the 'gold' hooks that Mustad export are in small sizes and go to anglers in the USA, France and Italy. It is known from history that hooks of pure gold have long been in use – and not only by rich men in ancient times and in the Middle Ages. The golden hooks of the Indians of Venezuela and Colombia are well known though this was before the Indians discovered what gold was worth. Pure gold is much too soft as a material, but the colour undoubtedly has the ability to attract. And more and more often one sees gilded fish hooks being recommended for angling competitions in fresh water.

A number of green hooks have gone to America, though it is doubtful if their popularity will increase. Whilst brown is constantly in demand everywhere, most of the blue hooks go to particular regions. Blue hooks are requested in Africa, in certain districts near the equator, and in the Far East. In recent years they have also gone to the USA and England.

On the world market, trout fishermen, as a general rule, prefer brown hooks, while the salmon fishermen traditionally stick to black.

Red is not encountered very often, except perhaps in France and Belgium. In the Nordic countries red is most often seen in connection with shrimp tackle when river fishing for salmon. Then the hooks merge with the shrimp's own red colour. As a matter of fact we are here touching upon something that will surprise quite a few: it is not true, as some people believe, that the regular sea shrimp, the trawl shrimp, is a greyish-yellow colour, and turns red only after it has been cooked. On the contrary, it has a bright red colour when it comes fresh from the sea, and only turns pale afterwards.

THE POINT, THE EYE AND THE SHANK

Is it still worth experimenting with unusual types of fish hooks? Or could we not accept that, with thousands of years of experience behind him, Man has finally arrived at the range of hooks that will cover every kind of fishing.

As a matter of fact, experiments are constantly being made. The future will show what the fishermen prefer, and no hook factory would wish to force its own viewpoints onto the consumer. On the other hand, there is considerable interest in recording all the experiments that are being made with different types of bends, points and shanks.

In Norway the preliminary tests have been carried out on the assumption that hooks with a curved shank will result in a better catch than hooks with a straight shank. The initiative was taken by fishery scientist Johs. Hamre, who explained his results in *Fiskets Gang* 1968, no. 46. Johs. Hamre experimented with two types of hooks, one with a straight shank (Mustad qual. 7296 B No. 6, in general use), the other with a curved shank, a new hook that had been made by Mustad at the request of the Department of Fisheries, and was marked qual. 7295 D no. 6. Two thousand four hundred of each type were used, and, after meticulously-recorded statistics, the conclusion was that the hooks with the curved shank appeared to provide considerably better catches than the hooks with the straight shank (18·6% better). This experiment was carried out with longlines in the waters east of the Shetlands. The dogfish was selected, among

79

other reasons, because dogfish hooks have a longer shank than usual. Thus, it may be assumed that the shape of the shank has a relatively greater significance.

Experiments have also been made with hooks in which the shank and the point have had different shapes. British scientists have apparently been amazed at the many hooks which have traditionally had a point with an exaggerated inward turn. We are reminded of the Indian hooks from Alaska (p. 17), and several of the bone hooks from the Stone Age. Scientists from the fisheries laboratory in Plymouth took as their starting point the so-called 'ruvettus-hooks' from the Pacific. The ruvettus is a fish that is found only in warmer regions; it is one of the escolars. The typical ruvettus hook has a point that turns more or less vertically against

The two types of hooks, somewhat reduced in size, that were used by Johs. Hamre during a longline experiment with spiny dogfish. The hook with the curved shank caught the most fish. (Illustrations from Fiskets Gang *No. 46, 1968.)*

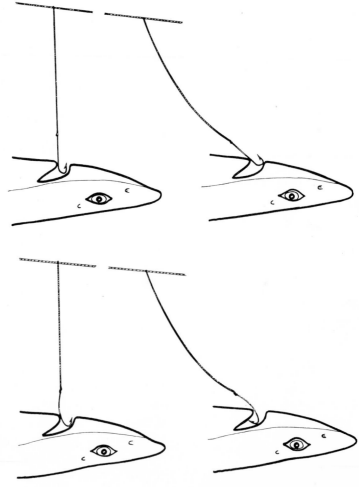

Right: Possible situations during longline fishing. Above: a hook with a straight shank that is proving a failure. Below: a curved hook that 'hooks'. The point of the theory appears to be that a curved hook will turn more easily to an advantageous position.

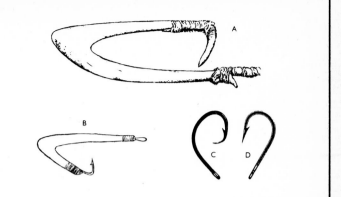

Fig. 1. A, 235 mm *Ruvettus* hook from Fox (1897). B, *Ruvettus* hook from Powell (1964); length *ca.* 50 mm. C, Size 6 Mustad Tuna Circle hook. D, Size 6/0 Mustad sea-master hook; length 51 mm, gap 20 mm.

the shank and extends an amazingly long way in towards it. Is it possible to fish well with this type of hook?

The British experiment is described in the 'Journal of the Marine Biological Association', 1973, no. 53. The fishing took place on the continental shelf in the Bay of Biscay, in waters ranging in depth from 800 to 3600 metres. The experiment was conducted during several periods, under the direction of G. R. Forster of the Plymouth Laboratory. Two types of hooks were selected, one the Mustad Sea Master, the other a Mustad Tuna Circle – the latter with a point that bent conspicuously inward. The British used relatively short lines, with some 10 metres between each hook, since this was a question of large fish in deep water.

Several species of shark were captured. For most of the bigger sharks the type of hook did not appear to have made very much difference. The smaller sharks, on the other hand, were captured more easily with the Circle hooks. But Circle hooks were especially effective when it came to fish belonging to the cod order. (*Mora moro* and *Antimora rostrata*).

Strangely enough, there have also been accounts from Australia about surprisingly large catches having been made with hooks in which the point is curved in towards the shaft at an exaggerated angle (*Outdoors*, Australian periodical, June, 1974).

It is probably too soon to arrive at any general

conclusions from such diverse experiments with hooks that have bent shanks or greatly curved points. But perhaps it is not solely a matter of chance that curved-in hooks appear currently to be gaining in popularity. For a long time, this was regarded as a 'typically American' point, but in recent years it appears as if 'curved-in' hooks are beginning to appeal to an ever-widening circle of fishermen.

Offset bends may be debated endlessly. As has been mentioned above, Mustad's fly factory has been concentrating on straight hooks. But what applies to fly-fishermen need not apply to a longline fisherman or one who fishes with a hand line. Straight points are customary, but, like the Icelanders, many swear by turned points for fishing with bait. (About recent Norwegian experiments see p. 135.) For ling and torsk the offbent hooks caught 20% better than straight hooks, for cod the difference was small.

But what about the turned bend – in which direction should it turn? It is pure conservatism that explains why most professional fishermen want the point to be 'kirbed', while sports fishermen largely prefer that it be turned the other way, i.e. 'reversed'. Perhaps it is mere coincidence that decides the choice. In certain regions along the coast of West Africa, near the estuary of a certain river, the professional fishermen demand 'reversed' hooks. But strangely enough, further north and further south of that same estuary they stick to 'kirbed'. The reason might be that the fishermen in the different areas are traditionally accustomed to holding their hands in certain positions while baiting their hooks.

HOOKS AND ANATOMY

Quite often it requires a zoologist to explain the need for so many different types of hooks since a part of the explanation rests with the fish itself.

If the fish has a tough mouth it may be an advantage to use hooks with a short barb and a short point. British anglers especially, and not without reason, have clung zealously to short points for salmon and trout.

In other cases short shanks might be an advantage, for instance if you wish the hook to be swallowed; this might be the case if you are catching fish with 'rotten jaws'. Long shanks might prove an advantage (for instance for flounders and dogfish) partly because fish with sharp teeth can cut the leader if the shank is too short, partly because it is easier to work with a long shank when loosening the hook from a powerful, wriggling fish.

In some cases the fisherman will save time by using hooks without a barb. In previous years, when there was a surplus of mackerel in the North Sea, the trolling fishermen often squeezed the barb flat, so that they could just shake the mackerel off the hook. They gained time that way, and if they lost a few more fish while hauling them in, this did not matter at a time when there was an abundance of mackerel.

Barbless hooks may also avoid damage to the fish. They make smaller wounds in the mouth of the fish, an important fact if the fish are so small that they ought to – or by law should – be put back again into the water. In order to protect the stock some countries have forbidden barbs on hooks for certain types of fish. The Swiss authorities have even considered barring treble hooks unless two of the three points are without barb.

As you can see from the illustrations on pp. 72–3 some of the hooks appeared to be abnormal – for instance those for catching pike, crocodiles and bonito.

The idea of the pike hook is to provide room for certain baits, usually a small fish like a roach. The point should be short because the pike can squeeze hard. One seldom loses a pike provided the hook has not caught in the bony part of the jaw.

The crocodile may be fished – or should we call it captured – in various ways. One of these occurs in Asia and Africa. The hook is lashed to the bait – a dead dog or what have you. The bait is fastened to a rope or chain that is made fast to a stake near the water's edge. Then they get hold of a live dog, and, by pulling its tail or in some other way, make it howl in pain. The crocodile crawls ashore, finds the bait, swallows it, and goes back into the water in order to digest its prey. One has to wait patiently, because it pays not to start hauling in too soon – not until the bait is partly digested. Only then can the hook stick in the crocodile's stomach.

The bonito hook is without barb. It may be used for trolling, but is particularly designed for fishing from a stationary vessel that carries live bait. This method of

Bonito trolling is a widespread sport. It can be lucrative as well, which is shown by this picture from Australian waters.

Opposite: Up to thirty men – in this case Spanish fishermen – may be working along the side of the ship when the bonito gathers. Notice the sprays of water that whip up the sea. The hooks are bright and without barbs. With short lines, strong poles and strong arms the fish are swung on board.

fishing is very old, having been used for centuries by natives in large areas of the Pacific. In our day this highly specialized method of fishing is carried on from larger boats with all the latest devices.

There is an old saying: 'One should never say never'. This is true of the barbless bonito hook. In the Pacific, some bonito hooks with exceedingly big barbs have been used. But, according to scientists (C. Nordhoff: *Notes on the offshore fishing of the Society Islands*, 1930) these have served the opposite purpose of most barbs: they prevent the hook from penetrating too deeply into

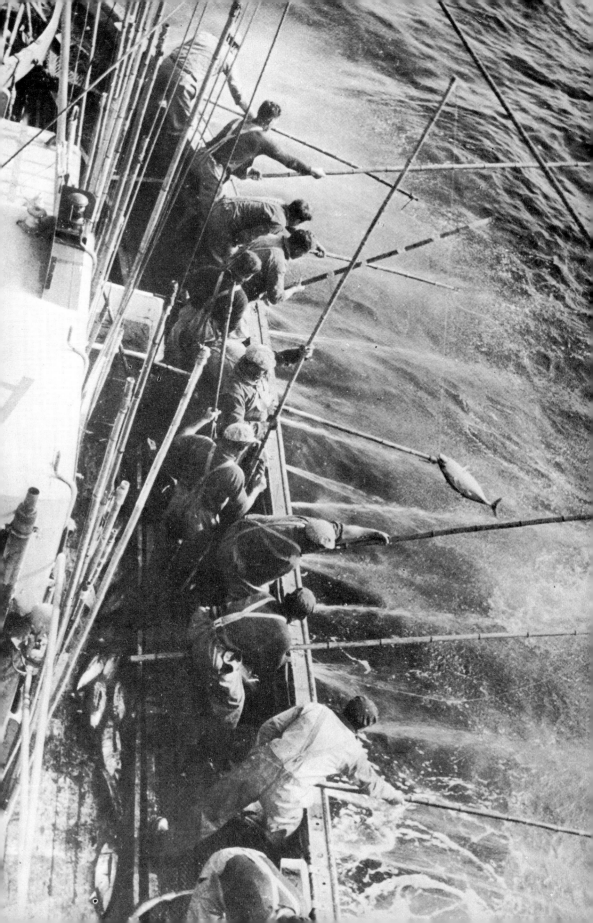

the fish. Every second counts when the bonitos gather alongside the boat. Then the barbless hook as well as the hook with an oversized barb can serve the same purpose – to get the fish off the hook quickly.

Today the typical bonito hook is constructed for a unique method of fishing, which is in use over a wide area.

The vessels carry a big load of small, live fish. In the vicinity of a bonito shoal the live bait is thrown out. At the same time water is sprayed out through a row of nozzles. The sprays of water resemble the splashing of fish when they hit the sea and make the bonito think that they are in the midst of a teeming shoal of fish. The crew stand by the rail, equipped with rods and lines of about equal length. With powerful tugs the catch are swung on board. If there are larger tuna, each line can be manned by two men who use two rods attached to the same line.

Year by year, treble hooks are becoming more popular. The number of treble hooks that are sent out on the market is increasing in proportion to the number of sport anglers. Or, to be more precise, it is increasing even faster, because there are also professional fishermen who are changing over to treble hooks for trolling and handlines. Of no little importance is the anglers' enormous use of casting plugs. Today all kinds of equipment – spoons, spinners, wobblers – are equipped with treble hooks.

Interestingly enough, it appears as if the treble hook is also making headway among discriminating fly fishermen. The Waddington, Edmund Drury and Tube Flies, are regularly equipped with treble hooks. In recent years even regular wet flies have been tied to treble hooks. There are anglers who emphatically maintain that treble hooks make the best catches, or, at any rate, hold the fish best. But here it is the word of one man against another. Some salmon anglers zealously maintain that flies on single hooks move most 'prettily' in the water. There are experienced salmon fishermen who claim that the salmon can wriggle off a double hook more easily than off a single hook. In short, every fisherman seems to have his own opinion. Much of the explanation is probably due to the fact that no two waters are alike, and the way a fly behaves depends on the weather, the wind and the current. Perhaps the type of hook will depend on whether the casting is done upstream or downstream. A treble hook is often heavier than a single hook, and so the choice may be on the basis of how high in the water one wants it to be.

In practice, the treble is a relatively uniform hook everywhere, with the exception, perhaps, that the French seem to prefer treble hooks with a slightly wider gap and a rounded bottom.

TENSILE STRENGTH

How pliable should a fish hook be? So far as the 'tensile strength' is concerned (which is the professional expression) then it must be said that no fish hook can satisfy absolutely everyone. A hook must not be too stiff, nor should it be too soft. It is practically impossible to find the one hook that will satisfy every fisherman under every condition.

The angler fishing in fresh water would certainly be disappointed if the hook should break. If the hook snags on the bottom and he has to use all his strength to get it loose, he would prefer that the leader broke, or the swivel, anything but the hook. He must be able to depend on the hook's ability to withstand maximum strain.

But let us take the example of the coastal angler with his casting rod. For some fish he wants to draw his plug as near the bottom as possible. Suddenly the plug gets stuck in stones and seaweed. He has lost many plugs through the years and has learnt that he can save the plug if the hook straightens out in the bend.

Nor does the longline fisherman off the coast and on the banks like hooks that are too stiff. The crew has no time to remove every single fish carefully from the hook. Thus the procedure is a bit brutal. On deck, the gaffer holds onto the fish until the gaff is jammed against the trawl reel, and held tight. Then, one of three things will happen: either the hook loosens from the fish without too much force, or the ganging breaks, or the hook will straighten itself out. The latter can be of advantage if the hook is fastened to some solid bone, for instance the jawbone of a cod.

So far as tensile strength is concerned, Mustad has tried to arrive at the right compromise, and believe they have succeeded in creating a suitable norm for hooks,

depending on their use. One difficulty has been that individual nations – particularly the Japanese, the French and the English – have preferred hard hooks. They would rather see the hook break than bend. This, however, is a preference that at long last appears to be on the wane.

The big game fisherman will seldom find a hook that is strong enough. But the price of hooks is a trifle in relation to what the rest of the equipment costs – the boat, rods, reels and everything else that is needed.

For this reason it has become a challenge to hook manufacturers to construct the best hooks imaginable for their big game customers – those who are battling with big fish like the marlin, tuna, tarpon, sailfish, shark and swordfish. It has been encouraging for Mustad to see that – in spite of intense international competition – their big game hooks have caught on. Here are two quotations from the literature on the subject:

In his book *Fishing in the Atlantic*, New York, 1947, S. K. Farrington Jr. writes:

The only hook I would ever use if I could possibly get it in any size is the Mustad. Both in strength and the ability to hook fish with consistence, it is by all odds the best hook ever made and on the market today.

From different waters comes a report that is equally complimentary: Athel d'Ombrain says in his book *Game fishing off the Australian Coast*, Sydney, 1957:

Several makes of hooks are available, but it would be safe to say that the majority of game fishermen use Mustad hooks. These are extremely strong and have very good points that are easy to keep sharp. I have never had one give the slightest trouble. Most good fishing stores stock them in a large range of sizes. These hooks are slightly offset, and are so nicely proportioned that it would be difficult to imagine a better hook to drive into a fish. Sizes 14/0 and 12/0 are mostly used for sharks on heavy tackles, 10/0 is large enough for marlin. Special longer-shanked hooks, size 16/0 are often used for sharks.

Right: *Strong equipment is needed for the shark. This picture is from Athel d'Ombrain's* Game Fishing off the Australian Coast, *and was taken just before this shark was gaffed. Mustad supplies shark hooks up to 35 cm. long.*

Below: *Piraiba (Latin Arapaima) is the local name of a river fish that, in South America, can weigh up to 350 kilos. The fish here weighed slightly less than half that. It was caught with a Mustad hook, qual. 1612, No. 12/0, over a white sandy bottom in the Araguaya River.*

THE SIZE NUMBER

Unfortunately there are no absolute measurements that can indicate the exact size of each hook.

Many misunderstandings could have been avoided if it were generally known that there are no absolute measurements for hooks. The assumption that there are appears practically everywhere in fishing literature. A typical example is the statement: 'Sizes 14/0 and 12/0 are mostly used for sharks'. It is never enough to designate a hook by such and such a number. Mustad has different types of hooks for shark, and what the number would mean depends entirely upon your choice of hook type.

In addition a number of other hook factories specify the size according to their own scales. The scales therefore may differ from factory to factory and from country to country. Some factories have based their scales on the length of the hook. Others, like Mustad, on the width of the gap, i.e. the distance between the point and the shank. The result is a regrettable jumble. Only the conclusion is clear and unambiguous: it is not much help to speak of a hook of such and such size – unless at the same time the name of the factory and the type of the hook (quality number) are given.

The deceptiveness of a size is apparent from the illustrations in the margin at the bottom of the page.

This example has been carefully chosen in order to prove the point. Never assume that a size number gives the absolute size of any fish hook! Indeed, it would be possible to increase the confusion of the reader even more. There are still varieties or brands of hooks in which size 1 is bigger than 1/0, and 12 consequently bigger than 11. There are not many cases of this, but they still exist on the Mustad export market – namely on older types of hooks, where the receiving country has insisted that the scale go in the opposite direction. Such reversed scales have not only been used in the Far East, but in former times also to some extent in Europe, Germany and England. In England the 'wrong' way was quite common in the past. However, it went out of fashion, and would probably have been forgotten had it not been for an able and ambitious fisherman by the name of H. Cholmondeley-Pennel, who published a book around 1890: *Modern Improvements in Fishing Tackle*. He had grown irritated with the confusion of numbers, and, with good reason, quoted a contributor

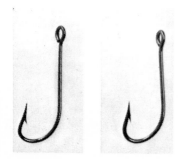

Both of these Mustad hooks are reproduced in their natural size. Right: a Best Kirby, Left: a Kirby Sea. The one hook (Best Kirby) is numbered 1/0, the other 12. Complete confusion! (Photo: Curt Ingemar)

The cost of his equipment is a minor consideration to the big game fisherman. This vessel fishes out of Hawaii, and from stem to stern it is custom-built for sportsmen wishing to pit themselves against shark, tuna, marlin, flying fish, swordfish, and the like.

The marlin is one of the liveliest creatures a sportsman can hope to hook. It makes violent jumps or 'walks' on the sea, steering with its tail while its body is above the surface of the water. In the picture the line is wound many times around its body – hence the bait-fishes seen alongside. In The Old Man and the Sea, Hemingway describes a struggle with a giant marlin. (Photo: Al Tetzlaff, California).

to the *Fishing Gazette*, who deplored 'the present stage of intolerable confusion that reigns in regard to the numbering of hooks'.

We learn from this book that 'the so-called Redditch scale' is not a scale at all, because it appears as if every hook maker in Redditch has introduced his own variations. According to the English angling expert, T. B. Thomas, the famous hook maker Samuel Alcock is said to have made a very sharp answer on one occasion when he was asked if it wasn't possible to standardize the methods of numbering of the Redditch factories. He replied: 'This is *my* scale!' And that was the end of that. As a result Mr Cholmondeley-Pennel added to the confusion by having a hook maker in Kendal (Hutchinson) introduce what he mistakenly called a 'new scale' – that started with 0/0/0 for the smallest hooks, while the largest hooks ended at 19. The result was complete chaos, and Mustad found it best to introduce their own scales, more or less in keeping with the average in Redditch.

It is quite another matter that Mustad, so far as possible, has tried to establish a uniform numbering of every single 'family' of hooks. In any case, the firm is proud of their painstaking accuracy in the numbering. Hooks belonging to the same series can be distinguished from one another with a difference of less than a millimetre in the length of the hook or the width of the gap. If one keeps to a familiar series, a given size will provide a guide to the size of the hook. So far as fly hooks are concerned, there are so few types, and the number of families is so small, that in everyday language misunderstanding need not arise.

The secret workshop

An aura of mystery has always surrounded the Mustad firm.

There are not many companies about which so little has been heard. No one has ever seen a stock market quotation for a Mustad share. On the whole, very few people know what the Mustad factories actually produce. What is known is that it manufactures a range of products differing greatly in variety and size. Some are weighed in grammes, others in tons. They range from hooks and pins to trawl doors and smokeless incinerators.

Of course, the *name* of Mustad is well-known. It may be seen on the legion of cartons, and labels for paper clips, drawing pins, zippers, various types of nails and screws and the like. But advertising aimed at the man in the street has been exceedingly rare.

There was only a single exception: when it produced margarine, the appeal was made directly to the public. Older people will remember the words 'Mustad's Margarine' painted in gigantic letters on the walls of a great many buildings, and there was scarcely a country store that did not have a red mailbox on which was painted 'Buy Mustad's Mono'. The factory had made the mail boxes itself, and distributed thousands of them free throughout the country.

Except for margarine the management has consistently played down the Mustad name. It is very likely no accident, but more a matter of principle, that it should have taken exactly 131 years before the factory at Gjövik was furnished with a nameplate. This intentional obscurity lasted until the firm became a stock company in 1970.

Little was made public of the company's geographical expansion during the nineteenth century. At regular intervals the firm expanded as an international undertaking. Mustad factories spread throughout Europe. Today the number would have been even greater had it not been for World War II. When it was over, Mustad factories in Yugoslavia, Poland, Czechoslovakia and Romania were nationalized by their governments.

In 1972, Mustad opened a fish hook factory in Singapore. Today this is the only Mustad factory outside Norway that produces hooks on a large scale. A couple of factories abroad were bought up for the reason that, among other articles, they also manufactured fish hooks. The factories that were purchased are still in operation, though not producing hooks. A factory in Auburn, New York, might be said to be an exception. The production of hooks was also discontinued there; but it now serves primarily as a warehouse and dispatching department for Mustad hooks.

From the very outset, the people in the firm were of an exceptional character. Originally, the name was not Gjøvik, but Brusveen in Vardal. The two municipalities of Vardal and Gjøvik were later merged. It all began in 1832, when two farmers, Ole Hund and Hans Skikkelstad, built the first factory – a two-storey, log building down by the Huns River. The motive power came from four water wheels.

'Brusveen Nail and Steel Wire Factory', as the firm was called, made steady progress. Ole Hund soon retired. Hans Skikkelstad and the man he eventually worked with – Sheriff Ole Mustad, his son-in-law – were hard workers. Both were periodically members of Parliament. Skikkelstad also became the mayor of Vardal, when the legislation governing town and parish councils was passed in 1837.

Two days after Skikkelstad's death in 1843, his obituary appeared in the 'Lillehammer Tilskuer' and 'Morgenbladet'. As people were frugal in those days, the announcement also announced that the firm had changed its name. From now on it was to be called 'O. Mustad, Vardal'.

This lasted a good thirty years, until 1 September,

1874, when Ole Mustad made his son Hans a partner. With that, the firm 'O. Mustad & Søn' saw the light of day. Hans Mustad was 37 years old when he became a joint owner.

In those days there were some 150 employees. Father and son were full of plans for the future. The premises were overcrowded and this ushered in an era of new plants and the buying up of other factories. From the middle of the 1870's considerable parts of the factory – machines as well as personnel – were gradually moved to Lysaker near Oslo where O. Mustad & Søn had purchased building sites, erected workmens' dwellings, and secured the rights to the waterfall in the Lysaker River. The deed to the purchase of the first plot of land is dated 1876. One by one, the nail factory, the horseshoe nail factory, the foundry and the axe factory moved there. Brusveen was given a free hand to diversify – for instance into fish hooks. The first hook machine started operating in November 1877.

Sheriff Ole Mustad was a man of vision. When he died in 1884, there can be no doubt that he had laid a good foundation for his son Hans to build upon. Ole Mustad's main aim was to make the factory as self-sufficient as possible. This was a bold idea in the middle of the nineteenth century, at a time when there were very few large-scale Norwegian industries. In the metal industry, the Kongsberg Armament Factory from 1812 is probably the only existing factory which is older than the Mustad firm. Iron and steel had to be purchased, but, otherwise, every link in the chain of production had to be controlled and kept within the gates of Brusveen. Most of the machinery was designed in the factory's own offices. The foundry delivered the materials that were needed for the construction of the machines, most of which were built in the factory's machine shop. The iron was rolled in Vardal, and the wire was largely drawn in their own wire drawer. The packaging was made on the spot and the labels were printed on the premises.

In those days Vardal was off the beaten track. Much of the iron had to be transported from Sweden by horse and sleigh in winter conditions, over difficult terrain; considerable unpleasantness and many angry letters resulted when the deliveries failed to arrive in time. Then

The beginnings were modest. When the first hook machine started operating in November 1877, it was in 'The Mill' by the bridge, the low wooden building with the tall chimney. The little house with two windows visible at the top of the picture in the background, was the office in those days. The packing department was an outhouse, so small that it is hidden by the office in this picture!

it was good to have a man like Ole Mustad living in Vardal. He became a prime mover in obtaining a town charter for Gjövik and a proper steamship wharf in order that the loads of iron and wire could come by boat across Lake Mjösa. 'Jern' in Norwegian means iron, and 'Jernbarden' was the first steamboat on Lake Mjösa. It was built at Mustad's expense in order to tow barges.

However, there were also advantages in being located in Vardal.

In the Toten/Vardal district, the manpower was the best imaginable. In this particular area people were accustomed to working with metal. There was a live tradition of every kind of metal work in the surrounding districts. Strangely enough both Redditch, in central England, and Gjövik, in central Norway, were small towns which became hook centres more or less because of the local population's understanding of metal working.

It was Mustad's good fortune that several of the employees were exceptionally gifted metal workers.

Ca 1896

98

One of these was Mathias Topp, the man who designed the hook machine.

Mathias Topp was from Vardal, and had been interested in things mechanical ever since he was a boy. At confirmation age he went from district to district repairing clocks. He himself made clocks of the type that are still known as 'Toten clocks'. In 1863 he obtained a position as a designer at Brusveen. At this time he had already made several inventions including a wool card machine. Engaged by Mustad, it was not long before he had designed machines – partly alone, and partly in collaboration with others – for the manufacture of such dissimilar things as boxes, shoemaker's pegs and hooks. The box machine was used at the Rödfos Match Factory, which today is the Raufoss Ammunition Factory. But most important was the machine that gave rise to the hook industry at Gjövik in 1877.

Topp built models of wood and cardboard and tried them out before the machine shop delivered the first ones in metal. The hook machine has been improved again and again, but its operating principle turned out to be so revolutionary that it has formed the basis of hook production right down to the present day.

The machine operated rapidly and with precision, and the firm's two leaders, Ole and Hans Mustad, soon agreed that, from now on, everyone at Brusveen would have to keep this under his hat. As far as possible, the design was to be kept secret, the slogan was to be: 'Act, not speak!' The vows of secrecy which the factory exacted from its machinists, are so rigid that they are unparalleled in any other industry – unless it be in the armament industry in wartime. The workers had to pledge – for all time – that they would not go to work in any other factory that was engaged in related trades.

'Not even my wife is to be admitted to the hook machines!' declared Hans Mustad. As a matter of fact, no stranger has ever been allowed inside that department. According to a reliable source, Christian Mustad – the only one of his five sons who was connected mostly with the hook factory – tried to enter the machine room and was thrown out by a recently hired watchman who did not know him by sight!

The principle of secrecy manifested itself in every conceivable way. When employees of the factory were to attend parties at Brusveen Farm, personnel from Department X were almost never invited if people from Department Y were there. Those who belonged to the sales department were not supposed to have anything to do with the production or accounting departments. It could happen, for example, that the salesman might hear figures about the production rates – figures that should certainly not be revealed.

In former years, the salesmen had an arrangement whereby they earned most on the products that were easiest to sell. This was probably why, after World War II, the firm decided that the same person should not sell both hooks and nails; originally, it was much easier to sell nails than hooks. They went further than this. The agents were not supposed to know each other at all, and the factory thus gave them cover names. Even the countries were given code names in order that no one could babble about which hooks went here or there. Brazil was called 'Leyser', England 'Lan'. The code name for the United States was 'Ikan', and for Canada 'Canikan'.

In addition, when new machines or machine parts were to be produced it was rare for any copy of the blue print to leave the director's office, so that the foreman had to carry the original with him. In case of extra valuable objects, additional steps were taken. No authentic drawings ever left the office. The employees were given measurements and a ruler, but the ruler would be specially made for the occasion – drawn with false centimetre measurements. It goes without saying that entrance to the central machine shop was forbidden. The employees made the parts of each machine, but the job of putting the parts together took place in isolation.

All this may seem exaggerated caution, but has the principle been worth the effort? In the course of the hook factory's 100 years' history, it has happened once or twice that employees have broken their vows of silence and taken jobs with competitors. One of the instances could have been very serious because the man in question was hired by a foreign hook factory. Thus, it was probably fortunate that this man did not know all the secrets.

The two-storey log building by the Huns River

where the production was once limited to wire and nails has today expanded into an industrial plant of impressive dimensions and considerable prosperity.

In some departments – particularly in the hook machine shop – the old 'No Admission' is still maintained. But, otherwise, visitors may wander about quite freely, for the very good reason that anyone who is not an expert will understand very little of the whole process. Most surprising to visitors is probably the fact that so much space is needed to operate a hook factory of Mustad's dimensions. As one goes from department to department, it becomes clear that the production is amazingly complicated, and that it encompasses an unexpectedly large number of processes.

The actual manufacturing process is easy enough to grasp, but there are so many links in the production chain that the visitor soon alters his views on many things. If he felt that fish hooks were expensive, he would soon have to change his mind. It begins to seem remarkable that a fish hook can be sold so cheaply.

Most of the hooks have acquired their final shape by the time they are conveyed out of the automatic machine department. From then on, it is the material alone that is to be processed.

The tempering process is most important, and absolutely essential to obtain maximum strength. The hooks coming from the hook machines are quite soft, but after they have been through the tempering furnace and have been quenched in oil, they are hard and brittle. Next comes a degreasing operation in which all traces of hardening oil are removed. Next in line is the process where the hard steel is drawn back to get that desired combination of hardness and flexibility which most people would recognize as 'spring steel'. The tempering process is completed by spot check hardening control to ensure good quality. Now they proceed to the polishing department, where they are tumbled smooth. After-wards the hooks go into different finishing departments depending on what kind of finish should be used, e.g. tin-plating, nickel-plating, gold-plating or various paint operations.

To be sure, there *are* secrets in the course of these processes, but they are secrets which can only be revealed in a chemical laboratory. The testing of new anti-rust methods is one of the pieces of experimental research carried out in the factory's laboratory. But the laboratory also has a supervisory function. The chemical composition of the steel is analysed, and the tensile strength is tested and measured. It goes without saying that the fluids in the factory's baths and containers undergo continual analysis and chemical control. In addition, there are special projects and a wide variety of random inquiries – caused, for example, by anglers who complain that hooks have broken.

This latter problem does not occur very often. If a hook breaks, it is seldom in the shank or eye, but in the bend – as a rule, just below the barb. If such a complaint occurs, it is usually a question of the fisherman's word against that of the factory. The angler is willing to take an oath that his hook has never 'brushed against anything', while the laboratory believes it can prove that the hook *has* hit something. The angler's fly attains its maximum speed at the very moment the forward cast begins and the hook is closest to the ground. When casting in rugged terrain, the hook could easily hit something during the forward cast, a twig for example. For a delicate trout fly under adverse conditions it may be enough that the hook has been in contact with the angler's own rod. The fisherman has barely noticed it, so he protests his innocence in good faith.

The factory knows conclusively that certain finishing processes are better than others with regard to corrosion. Formerly, at any rate, the blue colour afforded poorer protection than the black and was decidedly poorer than a coating of tin (or cadmium and tin). There is little point in the factory's giving advice, when there are customers who insist that a hook should look like this or that. Anyhow, it appears as if tin-coated hooks will be the subject of increasing interest in years to come. While they were largely used in salt water in the past, there have been more and more fishermen in recent years who are trying them in fresh water. The trend of the development depends not on the factory but on the anglers themselves.

A number of hooks are of rustproof steel and are marked 'stainless'. They are well protected against rust, as in the case of knives, razor blades, kitchen utensils and other objects that are marked 'stainless'. But only those

103

As we can see there are still links in the chain of production in which machines give way to people. Under certain circumstances, the packing of the hooks must be done by hand, or one may find a situation like this one. A matted pile of hooks is untangled with the help of the 'tangle stick', and then each hook has to be examined and counted before it finds its place in the box. This pile consists of Round Bent Hooks, 'ringed', and with a 'superior point'. (Photo: Frits Solvang)

hooks that are made of an alloy (largely nickel) can be guaranteed absolutely rustproof. The only trouble is that these hooks are expensive.

So now the manufacturing process is more or less completed. But the hooks still have a long way to go before they are ready for delivery.

Outside the packing department there is a long row of big wooden boxes. Each of them contains thousands of fish hooks. Most of the boxes have been provided with a green ticket. Green means: Ready. But a number of the boxes have red tickets. Red means: Stop! The red ticket indicates that the entire contents of the box must be emptied out and every single hook examined by hand, because these have not been approved by the quality inspection.

Examined by hand? As a matter of fact, only by scrutinizing every single hook can the boxes with red labels be saved. Every single hook must be looked at and the defective hooks sorted out, or else the entire box must end up on the scrap heap.

In all these wooden boxes the hooks lie in a tangled heap. Try to pick up a single hook; you will seldom succeed because the hooks are in a jumble, linked together at the bend or with the point of one stuck into the eye of another.

In the packing department the hooks are spread over the work tables in comparatively small portions. Here nimble fingers sort one hook from another while a mental count is made. Usually five hooks at a time are

Some imagination is needed to understand this picture. Compare these hooks with your own fingernail to get an idea of their size. The hooks in question are numbered 18. 'Powder hooks' like these are to be had in an even smaller size. The tinest hooks are delivered with either ring or flat. The hooks shown here have a flat. One thousand of the tiniest hooks on the market weigh four grammes! (Photo: Frits Solvang)

counted until there are exactly 100 hooks in each box.

All the hooks have to be counted, but, fortunately, this can now be done by machines, with the exception of some large and irregular types of hooks which must still be counted and packed by hand.

Most difficult for human fingers is the sorting and packing of the smallest hooks. In the packing department they are called 'powder hooks' because they are so tiny. Pick up some of them, shake them in your hand and then drop them again and you will discover that some of these hooks have become attached to your skin without your noticing it. The tiniest hooks on the commercial market are no. 28 of Series 540 and 94840 (Viking). A thousand of these weigh four grammes!

If it is difficult to handle such small hooks on the work bench, it can hardly be easier for the angler to make use of them. Whether the angler buys them flatted or with an eye, it must be quite a job to fix the leader on to the hook. A thread of synthetic fibre is seldom thin enough and will in any case be difficult to fasten.

Painstaking accuracy is demanded in the packing department. A box of 100 hooks must not contain 95, nor must it contain more than 100. The only solution to the problem is accuracy. Recent statistics prove that each box that is advertised as containing 100 hooks usually contains 100·3 hooks.

To conclude, some words about storage and delivery. An outsider visiting the Mustad stores might be tempted to propose some simplifications. Couldn't there be a section for cod hooks alone, for instance? All cod hooks assembled in one place, regardless of where in the world they are to go later on? But then he forgets the fisherman's temperament and his fidelity to old habits. Here is a good example. Years ago many countries took part in the longline fisheries on the Newfoundland banks. Many nationalities were fishing side by side – Canadians, Americans, Portuguese, Icelanders, Danes, Norwegians. But did they use the same hook? They all caught the same fish in the same way, but almost every nation preferred a different hook – and nearly all of them were Mustad hooks.

Many more complications arise. In a number of countries there is no demand for hooks of this or that type, for instance, 'kirbed' or 'reversed'. In England and

France there is a market for hooks that are only *slightly* reversed. In Cuba the demand is for the exact opposite – hooks that are slightly *more* offset than usual.

The fishermen's traditional conservatism is also revealed when boxes and packages are labelled. In certain regions the fishermen will not accept a box of hooks unless the label has gold lettering against a black background. It is essential for a box of Carlisle hooks to have a blue label, the sneck hooks a green, the O'Shaugnessy hooks a red.

At one time Mustad tried to introduce a lighter red colour on certain labels. They had in mind the dark nights and the poor lighting inside countless country stores. The dark red colours had the effect of rendering the boxes invisible. The idea did not work immediately; the new labels were regarded as a shipping error, and the boxes were returned. Then Mustad put two labels on the boxes – one a dark red and the other a lighter red. It took a long time for the fishermen to become accustomed to the lighter colour – and only then could Mustad use the light colour alone.

Many a strange tale could be told about the labels. In rare instances it can happen that the same hook is equipped with different labels, or the name is changed in accordance with local tradition. Most often it is the fishermen themselves who want it this way. They insist that such and such a hook be called this or that. They will not hear of any other name.

In spite of everything, it has been possible to carry out a degree of rationalization. Today the labels are printed only in English and Norwegian. Formerly they had to be printed in several languages. Not only has English gradually been accepted as a universal language, but also there has been greater acceptance of standard names. Thus in France, it is understood generally that a 'Hameçon Anglais' is the same as a Mustad Best Kirby, and a 'Hameçon Irlandais' is a Limerick.

Up to now Mustad has kept exclusively to the production of hooks. There have been very few exceptions. Many years ago the factory supplied jigs as well as some very special 'hooks' for squid and certain types of very simple spoons. Mustad still operates an extensive artificial fly-tying business in Singapore. But the principle was established when the manufacture

There is always a name for the spoon bait, but the name of the manufacturer of the hook often remains a secret. This photo of a pike is from Trench, A History of Angling, *and is said to have been taken in Sweden.*

of hooks began in 1877: the policy was that Mustad would not compete with its customers. Among the big buyers of hooks are, of course, the factories that manufacture fishing tackle. Numerous firms include Mustad hooks anonymously in their own equipment – in flies, spoons, jigs, spinners and wobblers.

Obviously there has been a well thought out plan behind the expansion of the hook factory at Gjøvik.

Our thoughts turn first to Hans Mustad, who must have been quite an exceptional personality, down to earth but imaginative, prudent but daring. He was a man in his prime – forty years old – when he and his father started the hook factory in 1877. On his father's death in 1884 Hans Mustad became the sole owner of the firm O. Mustad & Søn.

He had little formal schooling behind him. Most of

his general education was acquired from a tutor. His knowledge of conditions abroad is all the more amazing. What he knew of languages was learned mostly during his travels. He was familiar with most of Europe; he himself picked out the sites for new Mustad factories in the nineteenth century. With a view to inexpensive transportation and water power they were always built by a river.

At home, the entire district owed him a debt of gratitude. He not only expanded the factory, but also initiated projects which served the interests of the district. He helped build up the local banking system and had a considerable share in bringing the railway to Gjøvik. Thus, it was not only the employees of the factory, but also Gjøvik people from every walk of life who joined in the torchlight procession to Brusveen Farm when Hans Mustad turned seventy. And when he died in 1918, after having managed the firm for nearly half a century, it was as if a general had fallen. On the day of the funeral schools and shops were closed in Gjøvik.

Hans Mustad lived in an age of pioneering, when society needed men of his calibre: an autocratic patriarch with a highly developed sense of ethics and responsibility. In every mention of him, the recurring theme is that he was like a father to everyone, strict and just. '*He was very gruff and very straight*' as a 90-year-old worker once stated in an interview.

Heaven help the one who did not come to work on time. But he was equally strict with everyone. It is said that his son Halfdan had to come to work on the morning of his wedding, whether he liked it or not. Hans Mustad knew every single one of his employees down to the youngest. The children bowed to the ground when he drove past. He did not talk much, but knew how to make good use of his time. During his drives between Gjøvik and Brusveen he was constantly seen in discussions with someone. He turned the carriage into a conference room.

As has been mentioned, he had little formal schooling. So he was well aware of what a good education could mean, and he became a model for other directors when the factory started its own school for the employees' children. Brusveen became a cultural centre.

The two men who created the firm name 'O. Mustad & Søn': Left: Sheriff Ole Mustad (1810–84), and, by his side, his son Hans (1837–1918).

The firm landscaped its own well-kept grounds, 'The Schoolyard Park'. A library and a band were established, a bowling alley and a shooting gallery were built. The Rifle Team still exist under the name of 'Gjøvik and Brusveen Rifle Team'. Yes, even a dancing school was operated at the factory's expense. Considering the generally deplorable social conditions of the time one could speak of a 'workers aristocracy' at Brusveen. The factory's machine operators were called 'hook-men' in the local dialect, and it was an honour to be a 'hook-man' at Brusveen.

It appears that Hans Mustad had a natural flair for finding skilled co-workers, and deep friendship and relationships of trust developed between him and his co-workers. It is characteristic that in his will, Hans Mustad left instructions for the burial of inventor Mathias Topp. (Topp died in 1930, at the age of ninety, twelve years after Hans Mustad's death.) Mathias Topp's funeral was to be exactly like Hans Mustad's. The Gjøvik band, which had its start at Brusveen, marched at the head of the funeral procession. A gate of honour was erected at the entrance to the factory where the employees lined the route.

Hans Mustad was extremely temperamental, so it is no coincidence that the words 'loved and feared' are constantly used to describe him. A vivid account of his character appears below.

Before the turn of the century, the first three Mustad factories had been built in Finland, France and Sweden. A factory in England was to be next. It was a few years before World War I. As usual Hans Mustad had sent one of his foremen over to supervise the building operations. And this was to give rise to conflict, because . . . well, that becomes apparent in the following lines, written by an engineering consultant who was sometimes called in by Hans Mustad for special assignments:

Like all men of great personality, he (Hans Mustad) had his little idiosyncrasies, which could not be opposed without risk. For this reason, my friends often asked how 'Old Hans' and I got along, and I had every reason to reply: 'Excellently'. My admiration for his exceptional ability and integrity increased as time went by, and only once did we have a serious clash.

The incident concerned the new factory which Mustad & Søn was erecting in Portishead near Bristol and reveals Hans Mustad as the man of consequence that he was, in that he stood up for one of his old foremen.

Mustad had a number of practical men in his firm, men who had risen in the ranks from workers to entrusted positions as foremen and masters, and whom he regarded with a touching devotion. Most of these skilled and faithful co-workers were well-to-do men, since Hans Mustad, from time to time, gave them ample bonuses in recognition of their merits.

One of these faithful old workers was a man named Løkken, and he was in charge of the construction of the new plant in Portishead. It was not to be expected that this self-made man was especially well-versed in what we call modern technology, and it was here that the divergence arose.

I had proposed that an important section of the plant be built in a certain way. But word now came from Løkken that he could do this less expensively. I tried to convince Hans Mustad, with figures, that my way, the operating expenses would be so much less that the building costs would be recovered in a very short time. He did not like it – for the simple reason that he did not understand my technical estimates. Nor did he like the fact that his faithful old coworker should be defeated in a practical matter by a much younger man. He did not say this in so many words, but it was in the air. In the end, the tone of my voice no doubt revealed that I was a bit irritated, because Mustad suddenly said: 'Tell me, is this your factory or is it mine?' 'It is yours, of course,' I replied, 'but if you have hired me in order to protect your interests in this particular matter, then you ought to follow my advice.'

'I have been able to protect my interests myself during a very long life,' he said – and now his tone was a shade belligerent – 'So it shall be as my man suggests.'

'Very well,' I said, 'but then you can also manage without my further participation, and I will give this to you in writing.'

Then came the bombshell. Mustad drew himself up to his full height, pointed to the door and uttered a single word:

'Out!'

I went back to my office, and dictated a written confirmation, as I had promised. It was formulated in such a way as to leave no room for misunderstanding.

The following day, Mustad was on the phone and asked me to come down to his office as quickly as possible, as the matter was not suited to a written discussion. An hour later I was with him. Was I mistaken? Wasn't there a trace of humour in that otherwise austere face? Was it possible that he had even allowed himself to smile at my letter, filled as it was with youthful indignation? I do not know, but our meeting ended with Mustad giving me a free hand, and asking me to go to England and attend to it.

I was uneasy about my meeting with Løkken in Portishead, as I assumed that he was like most of the old diehards who overpopulated our industry at that time. Thus, I was quite pleasantly surprised to meet a nice, unassuming old fellow, who was easy to negotiate with, and the matter was settled to everyone's satisfaction.

The remainder of the story of the hook factory may be told in brief.

There has been an unbroken chain of expansion and modernization. Of special significance, perhaps, is the tradition that has been developing in the technical departments. Foremen and plant managers have often been succeeded by their sons, and this has guaranteed the company a unique continuity in special professional areas.

One of the older plant managers achieved considerable success when he managed to develop Mathias

Topp's hook machine for the automatic production of double hooks. A generation later, the son of this same plant manager contributed to speeding up the production of treble hooks. A treble hook is merely a double hook to which an ordinary, single hook has been added. New techniques have eliminated the previous method of soldering by hand in favour of brazing at an industrial tempo.

Neither Ole nor Hans Mustad had any particular confidence in scientists in the industry. They preferred practitioners to theoreticians. However, it should be added here that the firm had its secret workshop. The fact that the company wished to steer clear of engineers with higher education was probably a part of the strategy. It was most important to hire people who would receive their training in the factory and whose experience would be wholly within the workings of the Mustad plant. The 'outside experts' found it easy to change jobs according to any new offer that came their way.

The company had steered a cautious, but probably wise course in valuing practical knowledge more highly than theoretical training. One remark is worth remembering in this respect. It was uttered early in the 1920's on an occasion when scientific experts had been summoned to Gjøvik in order to examine an important stage of the hook-making process – the lacquering. None of the experts could find that the lacquering needed improving. 'What is science?' exclaimed one of the gentlemen. 'Science is accumulated experience!' The remark was made in tribute to the ordinary worker who has made himself an expert and understood how to utilize generations of accumulated experience. . . .

But times changed and a new age of science dawned. Then the situation changed rapidly. Again and again the hook factory was expanded and modernized, and new production secrets were added to the old. One example is the considerable sum invested by Mustad in induction heating for certain processes in the production of hooks. High frequency heating was still so new in industry – shortly after the Second World War – that the English firm that took care of the installation insisted on photographing the completed plant. The company refused point blank to permit it.

The desire to prevent important information from leaking out has also manifested itself in another, characteristic fashion: no patents have ever been applied for on anything related to the manufacture of hooks.

As early as 1935, a local history of Vardal described the hook factory at Brusveen as the biggest in the world. And this appears to be the position today.

To be sure, since 1935 there have been good years as well as bad. The occupation of Norway and five years of war, had far-reaching consequences. Foreign trade fell off completely, and it was equally disastrous that hook factories were started in several countries as a result of the decline in the export. This led to a series of lawsuits. In South America, as well as in other places boxes of hooks were discovered with labels that closely resembled (or were out-and-out copies) of Mustad labels. A naive reply was received from a Portuguese producer who had begun the manufacture of hooks during the war years. With reference to his agent in France, he said that it would be impossible to sell fish hooks unless Mustad's trademark was on the package. Gjøvik could have delivered great quantities of hooks to Germany and the German occupied zones during the war, but as little as possible was sent there, with the result that, among other things, the Germans also started to manufacture hooks. This turned out to be more difficult than had been imagined. The new hooks were not good enough, and the fishermen complained. It is known that the Nazi government contacted German experts in order to find out how to make hooks that would be more satisfactory to the fishermen.

After the war too, the hook factory at Gjøvik experienced certain difficulties. Currency restrictions or outright import embargoes were contributing factors. Sudden setbacks in certain countries were clearly attributed to government interference. In 1968, Mustad had a contract to deliver 93 million hooks to Burma: not a single order has been received since.

Even so, the rewards have outweighed the disappointments. In spite of considerable market fluctuations after the Second World War, the production of hooks at Gjøvik has risen at a speed that has exceeded every forecast.

Hans Mustad's five sons and, after them yet another

generation of Mustads – managed the firm until it was reorganized, and O. Mustad & Søn A/S was established in 1970. As early as 1905, Hans Mustad had taken all five of his sons into the firm. After 1918, when 'Old Hans' died, there were five owners. The need for a large top management is understandable when one takes into consideration the many varieties of the products and the large number of factories abroad. Each of the five brothers – Ole, Clarin, Halfdan, Wilhelm and Christian Mustad – has had his special field of activity. Their interests were diverse but, collectively, it may safely be said that they served well both the hook factory and the firm as a whole. What they did have in common was ambition, a good nose for business and an unusual degree of energy.

Hans Mustad's two eldest sons, Ole and Clarin Mustad, were both mechanical engineers. Ole, the oldest, was the persistent proponent of the idea of continuity, fully committed to his fathers business principles. Halfdan Mustad was the typical financier, internationally oriented on every question. Wilhelm was also a prime mover – his special field being margarine. Because he was involved in many plans for expansion he was known as 'the architect' among his brothers. Christian, the youngest son, was the purposeful salesman of the hook department. Ole Mustad left behind a note in which – with an eye to a future history (publicity that the other brothers strongly opposed) he begged to be excused from too much praise as an engineer. 'But,' he added, 'Clarin was a mechanical engineer of exceptional ability. It is hardly an exaggeration to call him a genius.' (In the Technical Museum in Oslo there are two automobiles that were constructed by Clarin Mustad – one, a single-seater, the other a six-wheeler. Both were built in Norway. The technical details were developed in the factory at Gjøvik. The four back wheels of the six-wheeler were all powered. There were also universal joints, springs and certain technicalities based on principles which were adopted by the automotive industry much later. He also built an early diesel engine). These were the years during and immediately after the First World War and it is known that the Mustad firm had completed plans to buy their way into the French automobile industry.

As an engineer Clarin Mustad was one of the active designers in the construction of a machine for automatic production of horseshoe nails. This machine was guarded just as carefully as the hook machine. One of the nail factories was in Rumania and during the First World War orders arrived from Oslo that the machines were to be destroyed if foreign troops approached the factory. This did not happen, but all of the nail machines were loaded onto barges in the Danube, ready to be sunk in the Black Sea.

The five brothers also inherited Hans Mustad's temperament. The conscientious Christian Mustad could flare up like a rocket. It is said that he blew his top when some Norwegian newspaper reported that Thoring's hook factory in Stavanger (later bought up by Mustad) had received a gold medal at a world exposition in Rio de Janeiro. 'Do you think it will harm us?' he telegraphed the salesman who happened to be in Brazil at that time. The salesman replied consolingly that medals of that kind did not mean very much. Direct contact with the fishermen would count for much more.

After a while this same salesman said he had found a far better way of getting publicity for Mustad in Brazil. One day he wrote home asking Mustad whether they would be kind enough to send him 'a nice looking hook, about half a metre long and shaped like our shark hook qual 4480'. The Rio fishermen, he added, would highly appreciate that gift; the hook did not need to be tempered, but had to carry the trade mark of Mustad.

What strange plan had he in mind? From primeval times the Rio fishermen had had a peculiar tradition. At the start of the fishing season they threw a hook, as big as possible, into the sea. This ceremony took place each year in June and the whole arrangement became a public festival. A great part of the fishing fleet, dressed overall, went out in procession to witness the ceremony not far from the shore. Afterwards there was dancing in the streets, music and a grand firework display on the beach.

When the next 'festa dos pescadores' (the fishermen's feast) took place in June, the Rio fishermen's organization was presented with the biggest hook ever made at Gjøvik. The Minister of Agriculture and Fisheries in

Ebisu, the god of Japanese fisher-men, who (according to Charles C. Trench, A History of Angling, *1974) is always depicted as fishing with a rod.*

person made a speech and threw the hook into the sea.

It is difficult to say why some customs disappear and some survive. It may be just a superstition, even a religious belief. If the latter, a producer of commercial goods like fish hooks will have to resist many temptations in the advertising field, for instance in France, where St Peter, the protector of all fishermen can be seen as a hanging mascot in cars belonging to anglers.

A proper hook god hardly exists. Only one, of modest character, is mentioned in a book by James Hornell, *Fishing in many Waters*, Cambridge, 1950. In the middle of India, near the river Cauveri, some 150 miles from the coast, Mr Hornell came across the shrine of a village god. The black, stonelike god resided in a small square building where red prints of the open hand were splashed on the stucco. Outside, on spikes driven into the wall at the side of the doorway, hung several monstrous double fishing hooks, 8–9 inches long, together with a large number of small hooks arranged around circular plaques.

I appealed to the villagers for an explanation . . . and learned that the god in this shrine has no distinctive name other than Thundil-Karar, 'the hook god'. He is the special protector of the fishing caste. . . . Chief of the duties of Thundil-Karar is the safeguarding of his devotees from attacks by crocodiles. These ugly saurians are plentiful in the river Cauveri nearby.

Unfortunately it has proved impossible to produce a photo of this singular hook god.

The fish hook and the future

Much of what the sea can yield in the future will never be caught with a hook.

One example is the krill, a tiny, shrimp-like creature which is the principle diet of the baleen whale and which is now considered as a possible food for humans. As is known, the stock of whales was drastically reduced in the interwar period. Russian scientists believe that the whales which were captured in the 1930's and just after World War II, must have consumed something like 25–30 million tons of krill a year. If all hope of building up a new stock of whales is abandoned, the krill could be one of the great, untapped food reserves of the earth. According to official statistics, 25–30 million tons represent nearly half of all the fish and shellfish that are caught throughout the world every year.

If, in the future, we choose to harvest from the sea the food on which the whale – naturally, and without cost to mankind – has grown big and fat, the catch will not be made with a hook.

As we shall see, a considerable amount of the wealth of the sea may still be harvested to advantage with a hook. Among important new factors is the spread of sports fishing, and the technical development of modern fishing with hooks.

Special conditions may also be worth mentioning. In the 1960's, when Norway was engaged in a development project in India to initiate Indian fishermen in modern fishing techniques, it turned out that the trawls and nets were being torn to shreds on the coral reefs. Far and away the best equipment that could be used in the coral waters for a large, abundant and nutritious fish of the perch family, was an old fashioned hand line and hook.

The fish that are caught by sports fishermen have little influence on world economy, and yet sports fishing is significant. In the 'Marine Fisheries Review', October, 1973 (a periodical published by the U.S. Department of Commerce) there is an item that provides food for thought. Robert W. Schoning, head of the National Fisheries Service, estimated that the share of the sports fishermen in American saltwater fishing, is close to two-thirds of the total catch made by Americans in sea water. Up to now, at any rate, the Americans have paid comparatively little attention to commercial fishing, but the statement appears to be sensational all the same. In all likelihood, the spread of the refrigerator and the deep freezer has been a contributory factor to sports fishing, more fish than before being used as food.

In Norway, we have seen how sports fishermen everywhere are making their way to rivers and lakes, and, along the coast, there is hardly an empty space. The trend is the same in most countries. The people in charge of the orders for hooks at Gjøvik claim they are able to read the economic trends throughout the world by the rise and fall in the demand for 'sports hooks'. One of the things revealed by the orders that come in, is the move to large cities. One asks why the use of sports hooks has been constantly increasing in Sao Paolo, for example, the answer being that there are four million more inhabitants than there were only ten years ago. In city after city the population is increasing rapidly. With city life comes a yearning to get 'back-to-nature'. And with money to spend – more than before, at any rate – fishing becomes the modern man's hobby.

In 1972–3, at the request of O. Mustad and Søn, and in collaboration with the factory, a Swedish firm conducted a market analysis in connection with sports fishing.

The result was a report which reveals that the United States is the number one nation of sports fishermen. The report concluded that only 25% of all American sports fishermen fished in salt waters. (Thus, Robert W. Schoning's estimate that sports fishermen took more out of the sea than professional fishermen, is all the more amazing). The report also pointed out that the number of women engaged in sports fishing is greatest in the United States where women comprise 27% of all sports fishermen, as opposed to 20% in Sweden and 7% in Great Britain.

In Europe, it appears that Sweden has the most sports fishermen in proportion to the population. No less than 19% of the Swedes make fishing a hobby. Next comes Finland with 16% of the population, then Norway with 15%, France 8%, and Denmark, The Netherlands and Great Britain with 6% each, Switzerland 4%, and Western Germany with 1·1%. Expressed in total numbers, this should mean that the largest number of sports fishermen in Europe are to be found in France. After the French come the British, with the Swedes a close third. It is, however, hardly necessary to add that such figures can only provide an approximate idea of the facts. Much depends, of course, on what is meant by a sports fisherman. The $1\frac{1}{2}$ million sports fishermen included in the Swedish figure are people who have fished at least four times a year. If one takes into account those who have fished twenty times a year, the number sinks to 600,000.

The Report estimated that in 1972 the angler spent on an average 80 Norwegian crowns for his equipment. Around a quarter of this amount went on the purchase of equipment directly attached to hooks: all kinds of spoons, flies and jigs.

The main trend in the development is clear and astonishing. In the course of a century or less a sport for the few developed into a pastime for nearly everyone.

Angling spread to the Nordic lands directly from Britain. It can be seen from detailed inventory lists of the landed gentry from the early eighteenth century, that English fishing equipment had already come into the possession of the rich (e.g. fine silver spoons for pike fishing). But socially no great changes took place. At the beginning of the nineteenth century angling in Scandinavia was still a very exclusive hobby. At that time one finds the first newspaper advertisements announcing that English rods and flies were on sale in the larger shops.

British anglers were fishing in Norway as early as the 1820's, and, from then on the names of a great many

anglers can be found in letters and literature. Norwegian angling writers seem to agree that Sir Humphrey Davy was the first known Englishman to try fly fishing in Norwegian rivers. (In 1829 he became an early anti-pollution spokesman, complaining that floating sawdust was spoiling his river fishing in several places). A great many British anglers followed during the next decades. These foreigners were welcomed but regarded with amusement, and certainly not regarded as pioneers of a new sport. To begin with they paid for their lodgings, but not for their fishing; the natives seem to have been thankful enough for their share of the salmon and trout caught. In Sweden, as late as 1853, the Swedish author F. B. C. Böster could write from Falkenberg in Halland:

The English used to come here during the summer. From the crack of dawn until late at night their only occupation has been to sit with their rods below the falls. In truth, a dubious pleasure for such a long journey.

Maybe it was not only the good fishing that attracted the English anglers. Norway's 'wild' nature – the deep fjords, the steep mountains, the vast areas of unspoilt countryside, the strange customs of the country people – it must all have had a strong appeal in those romantic days. A flood of books came from British anglers about their adventures in Norway. With scientific perseverance a well-known Norwegian angler, the engineer Trygve Berntsen, read through thirty-two of these books. He concluded that their authors must have been great explorers. No less than seventy-three of Norway's best salmon rivers had been tried and described in these books.

It became quite common to rent a river and a lodging for long periods, even up to ten years. Thus these anglers knew both their rivers and their salmon. A story is still told in Namdalen about Sir Henry Pottinger, who used to make his own flies. Once he caught a salmon that had attached to it one of Sir Henry's own flies. Some years earlier he had lost that salmon through a broken leader. Unfortunately the details of this strange happening are not known. Probably the hook was fixed not to the mouth of the salmon, but into a fin or the gill-opening.

Today there is neither space nor money enough to

England is the home of match fishing. Here we see a typical competition scene with the competitors spaced out at regular intervals. (Photo: Angling Times).

enable the public to indulge in salmon or trout fishing on a large scale. On the other hand there is a rapidly growing interest in fresh water fishing as a whole, particularly in competitions. Fishing competitions may be as old as the hills, but not until our day have they gained ground in fully organized form.

Great Britain is the home of the amateur competition. So, naturally enough, modern fishing contests began in England. During the nineteenth century, they spread from the industrial regions of central and northern England. The biggest events still take place in these areas. There are huge cash prizes, cups and medals. The annual Birmingham competitions include some 5,000 participants or more. In 1906 the first national championship in Britain was held on the banks of the Thames. It gave rise to strict rules involving what is permitted and prohibited. Only one line, one rod and one hook is permitted at a time; groundbaiting is permitted and you are allowed to change rod and hook, in which case, however, the bait must not be put on in advance, etc.

As time went on, international competitions were also held, and a world championship is arranged in Italy every year. But each country can only be represented by a single team of five members.

The 'matchman', in this instance a German (Gunther Weinecke, Hannover).

An angling competition in Japan. Artificial fish ponds have been constructed even in the heart of Tokyo. The object of the competition is sometimes a fish that measures only 4 cm, and the size of the hook is regulated accordingly. Often a single human hair is used as a leader. (Photo reproduced from v. Brandt, 'Fischfang').

The coastal areas cannot be emptied of fish as rapidly as rivers and lakes. In any case, it is apparent that sports fishermen are turning their eyes towards the sea more than ever before. Fishing in salt water is the sport that is developing most rapidly today.

The great ocean depths are also being exposed to scrutiny. Deep-sea fishing competitions are being

When whales drifted ashore, or fishermen caught a Greenland shark on their hooks, the myths of terrifying sea monsters flourished in popular belief.

Right: *We see one of these monsters, reproduced in 'De Nordiske Folks Historie' ('The History of the Nordic People'), printed for the first time in Latin, in Rome, 1555.*

The Author, Olaus Magnus Gothus, was the archbishop of Uppsala.

FYRTIOTREDJE KAPITLET

Om den norska hafsormens och andra ormars storlek

DE, SOM IDKA SJÖFART vid Norges kuster som köpmän eller fiskare, intyga enstämmigt en verkligen häpnadsväckande sak, nämligen att en orm af oerhörd kroppsstorlek, 200 fot lång eller ännu längre och 20 fot tjock, håller till i klyftor och hålor nära hafskusten vid Bergen. Denna orm beger sig ut ur grottorna endast under ljusa sommarnätter för att äta

A hook from the eighteenth century, about two feet long, forged by hand, for fishing Greenland shark in Troms, Norway. An iron chain, with free turning links, serves as ganging. The hook is baited with seal blubber.

(Illustration from priest Elling Rosted, 'Om haakiaerringen', Kgl. Norske Videnskabers Selskab, København, 1788.)

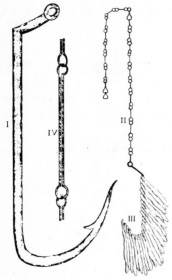

arranged more frequently and some enthusiasts are trying their hand at record catches in Norwegian waters. The fish in question is the Greenland shark, a very big species of shark which, in years to come, may attract foreign big game fishermen in Norway.

A Norwegian named Thorbjørn Tufte used Mustad's biggest shark hook when he wanted to pit his strength against the Greenland shark in 1974. His reel and rod were of the greatest possible dimensions. An enormous creature, obviously a Greenland shark, nibbled at the line. It was at a depth of 500 metres. The shark dragged out an additional 300 metres, but then it got away. Mr Tufte believes it is useless to have another try before an even bigger and better reel has been constructed.

On the hook he had a whole torsk and lots of seal blubber. The bait weighed 10 kilogrammes and the sinkers weighed the same. Big game fishermen who are interested, should know that a detailed description exists of fishing Greenland shark with a hand line. It was written by the minister of Tranøy in Troms, and was published in Copenhagen in 1788, but has been reprinted in the *Årbok for Senja*, 1974.

The minister himself had been out fishing for Greenland shark. It is easy to understand the belief in supernatural beings in the sea, when one reads of the struggles of the fishermen with the 'monsters' that could be pulled out of the waters in the Senja district. A fat Greenland shark represented a whole barrel of liver oil, and, in Bergen, more was paid for this than for a barrel of cod liver oil. The hook that was used was hand forged and two feet long. The best bait was the unborn seal pup or even the head of a seal complete with skin and whiskers. An iron chain with links that could be turned, was used as ganging. An ordinary chain that could not turn freely, would soon be cut.

We wish the big game fishermen 'good luck', but anyone who expects to establish a world record off the coast of Norway, is a bit too optimistic. A Greenland shark will hardly compare with the White shark that was caught with a rod in Australian waters in 1959. It weighed 2,624 pounds!

In addition to sports fishing the longline fisheries are attracting great attention today.

In recent years, there has been considerable coverage of longline fisheries on television. But the concept of longline fishing is still somewhat hazy for most city dwellers. The various techniques differ greatly, but common to them all is the fact that they use great quantities of hooks.

In the year 1715 English newspapers made a big fuss over a certain Captain Mulford, because he sewed fish hooks into his clothes. Those were the notorious days when London was infested with thieves and robbers.

Mr Mulford had come from the colonies in order to negotiate with the government about some unjust taxation on his homestead in Long Island. He became the hero of the day when he launched his new invention in the struggle against London's pickpockets. Should anyone try to rob him, they would inadvertently have torn their hands to shreds and, perhaps, even become hooked to him.

The same system was also employed in antiquity – this time in the Near East – for military purposes. Lines, to which great quantities of hooks were attached, were buried in the sand. Row after row of hooks were concealed where the enemy was expected. The charging soldiers stumbled in the lines, and their feet became entangled in the hooks. The system must have had the same effect as the mine fields in our own time.

The typical method of catching sturgeon has been with longlines, to which large, unbaited hooks have been attached upside down. A floater of cork or wood was fastened to the bend with a piece of string. Row after row of sturgeon hooks filled the estuaries of rivers in the Caspian Sea and the Black Sea. The same method was used from Siberia to Turkey. The hook pictured here is from the middle reaches of the Danube.

(Rohan-Csermak, Sturgeon Hooks of Eurasia, *Chicago, 1963.)*

It has turned out that fish may be caught according to this same principle. Baited hooks have not always been used in longline fishing, as is customary. The fish, in the dark of night or out of sheer curiosity, may become entangled in lines of hooks which have been strung out in rivers and lakes.

Even today, sturgeon is caught in this way – an extremely valuable fish which is coveted for its roe, the celebrated 'Russian caviar'. For centuries sturgeon fishermen have stretched their rows of longlines along the rivers and estuaries emptying into the Black Sea and Caspian Sea.

This brutal method of catching fish appears to be as unattractive as it is irrational. The Russians have prohibited it, one of the reasons being that hooks (as opposed to wide-meshed nets) will capture both large and small sturgeon, so the young fish will not have a chance of growing up. The Russians have persuaded almost all the sturgeon fishing nations – Bulgaria, Rumania, and the Iranians – to agree to the prohibition against the hooks. But not the Turks. In the estuaries of the Turkish rivers in Asia Minor they still use the old longline system for sturgeon. The hooks, which are generally forged by gypsies, are 20–40 cm long. The point is disproportionately long, without a barb, and is as sharp as a razor. More often than not, the hook is hung upside down, by means of a piece of cork fastened to the bottom of the bend with a piece of string. There is never any bait on the hooks. Two lines are usually fastened together at a time with approximately 130 hooks. How many sturgeons must be injured and die without being caught? When the spawning season begins in the spring, there is an average of 6–10 kg roe in an adult sturgeon, sufficient bait to catch the greedy and unscrupulous fisherman.

Elsewhere in the world, the general rule is that longline hooks are baited.

Most often the lines are stretched out horizontally. But the principle is the same if a line with a great many hooks is suspended more or less vertically in the water.

At times, the only practical equipment is the vertical line. One spring day in 1974, the author of this book was in Camara de Lobos, Madeira, looking at some local vertical lines. The staple food here is the black scabbard,

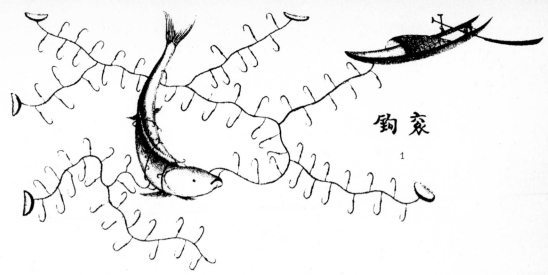

Ancient Chinese method of long-line fishing, still being used in Taiwan. There is no bait on the hooks, but the big fish (a species of sturgeon) become entangled in them.

(de Thiersant, 'La pisciculture et la péche de Chine', Paris, 1872).

an ugly fish, locally called 'espada', found at great depths. The fishermen use vertical lines that go down 1,400 metres. Each line has 160 hooks, and each boat carries four or five lines. Some of these are attached to buoys which drift freely in the water. There are no hooks for the first 600 metres. It all seemed quite primitive: most of the equipment was handmade, the line had been spun by hand and had been soaked in the juices of plants and urine in order better to withstand the salt water. Surprisingly enough, only the hooks were not homemade. They turned out to have been made in Gjøvik – regular Mustad Kirby Sea Hooks.

There are an infinite variety of horizontal longlines. There are lines that float and lines that lie on the sea bed. Other lines are suspended by floats near the bottom, at regular or irregular heights. Each leader ends in a baited hook.

The use of longlines is as common in fresh as in sea water. The lines may be set out around a lake, or in bends of a river. During the war, when food was scarce in Norway, longlines were used to such an extent and so effectively that longline fishing in fresh water is now prohibited by law in a number of areas, thereby putting an end to centuries-long traditions. Relics of the past are to be found in museums, where the finest craftmanship has been bestowed upon chests in which the lines were stored, with the hooks suspended neatly side by side. Along the shores of a number of Norwegian lakes, like Mjøsa, ancient 'bait cellars' may still be found, wells

*The black scabbard has a formid-
able mouth. It is especially plentiful along the coast of Madeira and is
caught with vertical lines at a depth
of some 1,000 metres.*

which were dug by the farmers in order to keep their
bait-fish alive.

So also along the coast. In old boat houses stood big,
heavy stones called 'angle whetstones', whose deep,
narrow grooves reveal that generation after generation
have sharpened their hooks on them.

Remarkably enough, longlines have also been
prohibited in salt water. Professional fishermen have
always been wary of newfangled ideas and when
Norwegian longline fishing began – probably in the
sixteenth century – the fishermen, who used hand lines
in the oldfashioned way, protested that wherever
newcomers set out their expensive longlines, the half
dead fish on the lines scared the other fish away. So
many complaints came in to the Bishop and the King,
that longline fishing had to be prohibited by law. From
1627 down to 1816, the use of longlines in the sea was

prohibited. During the nineteenth century longline fisheries mushroomed in Western Norway, not least in Lofoten. The longline fisheries assumed such proportions that they obviously became a factor which influenced the decision of Mustad to begin production of hooks in 1877.

It is impossible to say precisely how many hooks a longline fisherman needs. It is no small number, at any rate. Before the fishing vessels were motorized, a regular coastal fisherman kept mostly to modest lines that had from 100 to 250 hooks. But as a rule, he fastened several lines together. When fishing for cod, coalfish, haddock, torsk, ling, etc. he usually had a fathom or so between each hook. On the halibut lines, it was approximately 10 fathoms, and a very large bait was needed, for example a whole haddock on each hook.

During the early days of Mustad hooks, fishing vessels were equipped with decks so they could sail from Norway to far-off fishing grounds such as Ireland and Iceland. We probably learned most from the British,

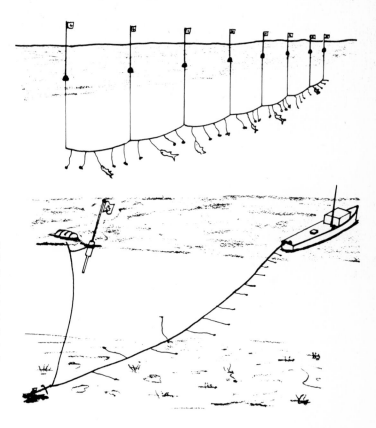

There are many different methods of longline fishing. Here are two examples: Above: a floating line in which the hooks are hanging close to the surface. Below: a line that is anchored and is being set out by a moving boat. Depending on the system the hooks remain on the bottom or are kept up by floats near the bottom at even or uneven levels.
(Illustration from Commercial Fishing Methods, *London, 1971.)*

131

A floating line being hauled in.
These are Norwegian fishermen
who have been fishing with longlines for salmon. The photograph
was taken some time ago when this
kind of fishing was still permitted.
(Nils Arnold Svaleng)

who conducted extensive longline fisheries for cod,
haddock etc. They started early with big, specially
constructed longlines. Salt water was kept running
through tanks in order to keep the catch alive. Each of
these longliners was equipped with 8–10 lines, each of
them had up to 5,000 hooks.

With the change to steam new technical develop-
ments followed. Today, the largest Norwegian longlines for whitefish in the Barents Sea can have a length up
to 40 kilometres (lines joined together). When the
longliners set out from Western Norway on a three week
voyage to the fishing banks in areas like the Shetlands,
the Faroe Islands and the Hebrides, each boat usually has
some 25,000 reserve hooks, in addition to the hooks
already on the lines.

Japan is the world's biggest fishing country, and they

also have the longest lines in the world. Their tuna lines can be 150 kilometres long and stretch over vast areas from the Pacific to the Atlantic. Of course these lines consist of sections joined together. Each section is usually three hundred metres long. Altogether these longlines may contain some 2,000 tuna fish hooks. The hooks hang at various depths, because the lines dip down to a depth of approximately 200 metres. A whole fleet of boats is needed – large and small vessels, factory ships and light transport vessels accompany the tuna longliners.

In 1974, according to FAO estimates, 70–75% of the world catch of 'tuna and related species' was made with hooks. However, in the case of tuna a certain percentage must be deducted. The expression 'related species of fish' includes varieties like swordfish, flying fish and marlin, of which pretty well 100% are caught with hooks.

Every so often 'longline fisheries' suddenly become the centre of public interest. Had it not been for the fact that it involved conflicts between nations and gave rise to protests, few people would have known that the Danes were engaged in large-scale longline salmon fisheries in the North Atlantic.

From time to time hooks loosened from a longline, have been found in the mouths of salmon caught in Norwegian fjords and rivers. 'Typical Danish hooks!' maintained indignant Norwegians – despite the fact that the Danes were using Mustad hooks.

The eel may also be the object of longline fisheries. Eel hooks resembling gorges were mentioned at the beginning of this book and present-day gorge-type hooks go mainly to France. The Netherlands, Belgium and Germany, use mostly regular hooks with a bend. Along the German coast of the Baltic the eel lines are 3–5 kilometres long, containing some 2–3000 hooks.

During the twentieth century basic revolutions have taken place in most fisheries. Longline fishing has generally been abandoned in the competition with more lucrative methods of fishing.

The steamship superseded the sailing boat, and the motor vessel superseded the steamship. With more room on board for bigger and heavier equipment, and with new manoeuvrability and speed, the fishermen thought up new methods of catching fish. Shortly after the turn of the century, purse seine fishing began in Norwegian waters. In the 1930's, the trawl became popular. As far as nets were concerned, new inventions and improvements followed thick and fast. Synthetic fibres were stronger, less visible and did not rot. The fisherman no longer had to sit half the day mending his nets.

But this is the way it has been since the dawn of time. Archaeologists tell us that in Mesopotamia and Ancient Egypt, for example, fishing with hooks was neglected for long periods of time – apparently because other equipment like fish traps and nets proved to be better. But sooner or later, the fishermen returned to the hook.

When Mustad made their first hooks a hundred years ago, it was an open question whether the longline fishermen in the north of Norway could be persuaded to buy them. Will history repeat itself? The question now has quite a different significance, because this time it involves the entire future of longline fishing. After many years of dormancy, it is quite possible that longline fishing is about to make a comeback.

Two main factors must now be taken into consideration. The first is higher oil prices, the other is the technical development of the fisheries.

A tremendous amount of horsepower is needed in order to drag a trawl through the sea. The skipper of a trawler can be content if he manages to catch two kilograms of fish for every litre of oil he uses. The fuel consumption of a longline fisherman would be quite modest by comparison. In addition, trawls and seines are expensive equipment, and their synthetic fibres are derived from oil products. Rising oil prices will mean increasingly more expensive equipment.

In addition, the use of trawls and seines presupposes a concentration of fish in the sea. By comparision a longline vessel can operate more effectively when the fish are scattered. When the day arrives that the price of oil goes even higher, and if the stock of fish is thinned out substantially, it may be difficult to operate lucratively with trawls and seines.

A longline will never be able to catch all the fish in a shoal, nor does it scrape the bottom, or damage the smaller fish. The longline also results in a catch of the best quality. Fish that are caught with hooks are almost always fresh when they come out of the sea. If they are caught in a trawl net or a seine, it is inevitable that a certain percentage will be damaged.

Thus, in recent years, it is not surprising that considerable efforts have gone into attempts to rationalize longline fishing to make it more competitive. It is about time too, because, as against other methods of catching fish, fishing with hooks has been a neglected area from the point of view of technical improvements.

It is no coincidence, then, that longline fishing in European waters only recently became the object of scientific research. In Norway, up to now, this research has mainly been concentrated on the catch efficiency of different gear parameters. In 1975 and 1976 The Institute of Fishery Technology Research in Bergen began some thorough research that might interest professional fishermen in many countries.

Part of these Norwegian investigations have to do with the shape of hooks. For ling and torsk offset hooks (kirbed) have been proved to be about 20% more efficient than straight hooks.

Even more interesting is the research into the materials used in the lines and leaders themselves. It has now been possible to demonstrate that monofilament longlines with nylon ganging fixed to the line with swivels, can catch much more than usual lines and ganging made of hemp, cotton or artificial fibres. One should add, however, that the distinct increase in the potential catch due to the use of monofilament with free hanging leaders is greater when the fish are less willing to bite – for example, when they have had plenty to eat over a long period (as with the cod in north Norway at the end of the chapelin season).

Is it possible to mechanize fishing with hooks?

For deep-fishing with a handline there appears to be a fair chance of success. Boat after boat is now being equipped with small 'machines' that operate the jig from the gunwales of the boat without the use of muscle power. The Japanese have even constructed machines to replace the fishing rod in bonito fishing. The bonito, a smaller species of tuna, are automatically flipped over the rail!

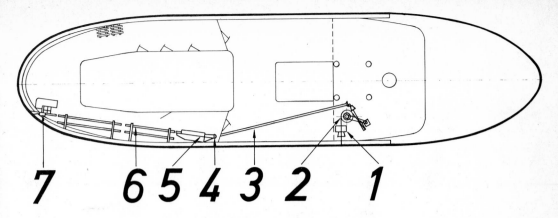

7 6 5 4 3 2 1

A Norwegian vessel equipped for fishing on the banks with Mustad's Autoline System. The sketch shows the most important links in the process after the fish have been gaffed and the line is being hauled on board:

(1) The line rollers are followed by the hook cleanser which removes the remains of bait and unwanted small fish.

(2) Line winch that hauls in the line.

(3) Guide tube for line.

(4) 'Twist remover' that keeps the gangings clear of the line.

(5) 'Hook separator' in which the hooks are intercepted and carried to the magazine rail.

(6) Magazine rail, where a manual repair of the equipment is possible.

(7) Baiting machine, which cuts up whole bait fish and baits hooks automatically as the vessel travels at full speed.

In this picture of Japanese bonito fishing ancient and modern methods compete. Standing in the background are fishermen who catch the bonito traditionally, by hand. In the foreground we see four fishing rods that are untouched by human hand. Machines lift the rods and flip the fish into the boat. The white strip of spray along the side of the ship deludes one into believing that the vessel is moving at high speed, but if we compare this picture with the one on page 85 we see that this is an optical illusion: the artificial sprays of water are whipping up the sea.

On quite a different scale are the experiments which, in recent years, have been carried out in several countries – experiments which aim at the complete automation of longline fishing. The idea is a daring one, but it can be realized. In fact, it has already been put into practise!

In Norway, during the 1960's, new inventions and ideas connected with the modernization of longline fishing appeared. The Directorate of Fisheries displayed considerable interest and provided support. Mustad concentrated on a project that had been worked out by Konrad Harem, a skipper from Möre. The construction was simple and sturdy. (Interestingly enough it was fishermen from Haram municipality who, in the 1920's, carried out the first experiments with trawl nets that were especially suited to Norwegian conditions). Mustad later blazed their own trail and developed – partly in collaboration with another Norwegian firm – the first 'Autoline', which was launched in the summer of 1971.

Like so many inventions, this one also had a difficult birth. The problems were largely connected with the link in the process where the baiting takes place. Mustad was forced to find new methods. With a grant from the Norwegian Research Fund of the Fishing Industry, the first completely successful tests were carried out on the Shetland Banks during the summer of 1974.

The Mustad 'Autoline' presupposes vessels of a certain size. Complete automation means that the fishing may be conducted with fewer men on board. The work proceeds more rapidly so that the effective fishing time is prolonged. Alongside the boat the fish are

Above: *A glimpse of the storage rail in Mustad's Autoline.*

Times change! In the old days, it took half the day to get the longlines ready for use. The picture on the right is a Wilse photo of the late twenties and shows fishermen from Lofoten engaged in the arduous task of cleansing the hooks and putting on new bait, so the line lay ready for use in the tub.

Today, automation has made it possible to do this work at very high speed. The Autoline machine baits four hooks a second, i.e. the baiting takes place on board at the same time as the line is set from the vessel at full speed.

gaffed manually. But from that moment almost the entire process is carried out mechanically. The hooks are automatically cleansed of the remains of fish and bait, and then baited anew before the line goes out again. It turns out that Mustad's new baiting machine operates with amazing speed, reliability and precision. During the decisive tests on the Shetland Banks in 1974 the machine could bait up to four hooks a second. Alternately hand- and machine-baited lines were set out and it was proved that the new Autoline system was highly effective. A saving of time represents increased earnings per man. In addition, the work on board becomes more attractive than before. The fishermen have always had a specially demanding job on the longline boats.

The time is long since past when Mustad felt they could manage without the help of science. Especially in connection with the development of the Autoline there has been the need for contacts, not only in the practical fisheries but also among those who are carrying on fishery research.

Is it conceivable that artificial bait will be used in the fully automatic longline fisheries of the future? If so, there are a considerable number of questions to which science ought to be able to provide the answers: How does a fish bite? What causes a fish to swallow a hook? What, if anything, is known about the various behaviour patterns of the different species of fish?

On a number of occasions the Institute of Marine Research, in Bergen, has kindly placed equipment, personnel, aquarium and research tank at Mustad's disposal for experiments.

One of the first tests was the filming of cod biting.

The first impression of the films was disappointing. The procedure was quite simple: a line with a hook and bait (frozen herring) was lowered into the aquarium-tank and held motionless a short distance from the bottom. Seven different hooks were used for the 'fishing'. Unfortunately, all the hooks were of a similar type, with a straight shank, and the outcome was as could be expected. It did not appear to make the slightest difference to the aquarium fish that the hooks were not all quite the same.

The fish were interested immediately. Almost all of them went to have a look at the bait, due presumably to appetite or curiosity. The reason the 'fisherman' kept his line still during the filming, was that the film was intended to help longline fishing, in which the hook is a 'passive' implement. It was a surprise to see that the fish constantly took the bait but did not swallow it. They kept the baited hook in their mouths – some for only a moment, others from 8 to 10 seconds – before they decided to spit it out again.

Perhaps it could have been expected that filming of fish in an aquarium would be of little value; it hardly comes as a surprise that bored fish in a glass tank lose their appetite and at the same time, have their curiosity roused when something new happens. But several scientists in recent years have noticed that cod often spit out a bait that is motionless. In Scotland behavioural scientist Anthony Hawkins has filmed biting cod in aquariums as well as in a free state. His experience is that

the biting cod behaves in the same way whether free or in captivity. A cod can keep bait in its mouth without swallowing it for a remarkably long time. Mr Hawkins took his underwater photographs at a depth of 10–20 metres. His aquarium pictures are so much clearer, that they are to be preferred for reproduction. The two that are reproduced here, are from the aquarium, but as far as the interpretation of behaviour patterns are concerned, Mr Hawkins assures us that they are identical with the photographs from the sea.

Mr Hawkins explains:

The feeding behaviour of a fish like cod, is rather different to our own. The cod seems to like to take the food into its mouth to test it. First the cod swims around the bait, trailing its pelvic fins and barbel (both organs containing taste buds), and perhaps touching the bait with these. Then, it ingests the bait, and tastes it with the inside of the mouth, and the tongue. This phase may take up to several minutes. If the fish decides to take the bait, it then pushes it back into the stomach using the pharyngeal teeth – hard rough pads at the end of the throat – or it rejects the bait by spitting it out.

My observations suggest that the fish is not generally hooked while 'sampling' the bait, but usually when the bait is being pushed into the stomach by the pharyngeal teeth. A very high proportion of fish caught on longlines are hooked behind the pharyngeal teeth – that is after they have swallowed the hook. This is why it is so easy to catch cod on a gorge.

What conclusions may be drawn from this? Complicating factors intervene. A fish has more senses than smell and taste. It can see. And everything that moves and appears to be alive can stimulate the urge to bite. When a cod bites at a jig, it has nothing to do with smell or taste.

If you have ever sat in a boat, fishing with a line and bait, you must have had the following experience: the line has been out for a long time, and you think you must pull it up now because the bait is probably gone. You start to pull and at the same moment there is a fish on the hook. In that case, it is not unlikely that the fish had the bait in its mouth, and was hooked at the moment you started to pull. But the explanation may also be that the fish bit because it saw something that moved. The casting angler has learned this lesson, too. It is not for

The cod's way of biting, studied at the Marine Laboratory, Aberdeen. Above: A cod sizing up its prey. It has extended its barbel – an organ of touch – in order to determine whether the bait is worth taking.

Bottom: A cod which has taken the bait and the hook, but hesitates to make up its mind – to swallow or spit it out. Anthony Hawkins, who has conducted a number of behavioural experiments, believes that it does not make the slightest difference whether the leader is thin or as thick as the one in the picture. He maintains that the use of 'invisible' leaders is of no advantage to the cod fisherman.

As far as this latter question is concerned, it should be added that opinions differ greatly. The Norwegian longline fishermen at Lofoten maintain quite emphatically that a fine leader results in the best catch.

(British Crown copyright, reproduced by permission of the Controller of Her Majesty's Stationery Office.)

nothing he has learnt to reel in with jerks. Or let us think of a group of holidaymakers who have never fished before. One sees that some catch more than others, even though they are all sitting in the same boat. It is conceivable that these 'lucky people' have more sensitive hands, are quicker at drawing in the line, or, with continual activity, 'bring the bait to life'.

The films that have been made clearly reveal that cod are stimulated by movement. On rare occasions, when the 'fishermen' in the aquarium tugged at the line, the cod would bite even though there was no bait on the hook.

Now and then, the 'fishermen' saw to it that the cod really was hooked. Then one made a surprising observation. Even though the fish wriggled violently as soon as the line was drawn up, the other fish that were free, appeared to be completely indifferent to their companion's struggle for its life. This tempts one to believe that cod lack a system of communication with one another to reveal that there is danger about. There is also reason to believe that the cod has a poor memory. Again and again the same fish took the bait it had just spat out. It remains to be emphasized, however, that experiments of this sort are still incomplete. Scientists maintain that cod are among the varieties of fish that are comparatively easy to train and are quick to learn. There are even scientists who do not believe it unlikely that, in future, the cod may learn to keep away from a trawl. This heavy gear creates a commotion in the sea, as it sweeps along it causes vibrations which the fish both hear and feel before they get caught. Sound, as we know, is conducted more rapidly and easily through water than through air.

Man's knowledge is impressive, as far as the fish itself is concerned – its build and functions. The fish's sensory organs and the way they work, have largely been explained from a purely objective point of view. Less is known about the use of the senses, and far too little is known about the reaction of fish at the sensory level. In questions concerning the behaviour patterns of fish, science appears to be on the threshold of knowledge. 'In this sphere, we still find ourselves in the Stone Age,' as one scientist puts it.

'Among the things that have astonished me,' said

Hilmar Kristjonsson, Senior Fisheries Officer in the FAO, 'is the way my childhood beliefs about cod have been proved wrong through the years. As a boy I always thought that the cod was timid, quick to slip away at the moment of danger. But my experience in later years has been just the opposite. If one puts a stick in the water, a cod will not necessarily flee. On the contrary, it might attack instead. A coalfish will react quite differently; it will flee.' If this assertion is correct, it should indicate that the cod will *not* attempt to escape an approaching trawl at any cost; it might even go to meet it instead!

As yet, too little is known. The question is of far-reaching significance without doubt – not only for the great fisheries, but also for the person who is fishing from a boat with the motor running. Will the sound of the motor, or the spinning of the propeller, frighten or attract the fish? It may be safely said that various species of fish react differently. According to prevailing knowledge it is possible that low frequency vibrations – for example vibrations in the water caused by the sound of a motor – may stimulate cod, while they frighten herring.

Work on artificial bait, as a goal for the future, is now in progress. The advantages of a practical artificial bait are obvious. An automatic baiting machine would be guaranteed 100% effective, and it should be possible to manufacture artificial bait far more inexpensively than natural bait. The fact is – at least as far as Norwegian fishermen are concerned – that it is becoming increasingly difficult and expensive to obtain good, natural bait. The best types of bait for cod/haddock and ling/torsk are herring, mackerel and squid, precisely the three varieties that have become scarce in Norwegian waters. Longline fisheries require considerable amounts of bait; up to 400 kg. frozen herring are needed in order to catch 5–6 tons of cod.

But how could one produce artificial bait that is good enough to compete with natural bait? The scientists have been given a hard nut to crack. In the first place: which qualities of the bait attract the fish? Is it the smell, the movement, the taste or what? Perhaps, even, the sound?

Scientists in a number of countries have been experimenting with sound. One of the Japanese projects has been to devise a sound that will scare away sharks but attract tunafish. Radio transmitters have been attached to tunafish, and their movements have been followed – not only their migrations, but also their reactions to noise. Perhaps it will not be long before British investigators will submit the results of studies, which have been in progress for years, about the reactions of our northern species of fish to sounds. The fish themselves emit noises, and it may be assumed that these are received, because they are all within the range of frequencies the fish command. If a bird can be attracted by a birdcall, why shouldn't a fish be attracted by the 'voice' of a fellow?

The longline has been called a passive implement. But is the hook on the line never in motion? Yes, it moves as the line is being set out, and it should be remembered that the adjacent hooks are set in motion as soon as the fish bites. However, any fisherman – not least an angler – will know that it is not only a question of what the implement looks like, but also the way in which it moves.

Does the fish have poor vision? According to an old saying 'fish are nearsighted'. But this is no longer the accepted view. According to modern science, most fish are not nearsighted, but have weak eyesight – at least in comparison with human eyesight. The fish sees just about what it needs to see, as a rule about as far as the water is clear. The eye of the fish has a far greater range of vision than the human eye. Unlike man, the fish does not have to change the position of its eyes in order to see in another direction. This might indicate that, in general, it is difficult for fish to see a motionless object and conversely that it would register, and react to, anything that moves.

When it comes to longline fishing, in which the bait largely remains motionless or lies on the bottom, it may be assumed that the bait must give off an odour in order to stimulate the fish to bite.

In collaboration with Mustad, the Ministry of Fisheries in Bergen has been engaged in a research project for years, in order to determine, if possible, the reactions of fish to smell. This, however, is a long-lasting project, and at the present time it is too early to publish the results of certain basic and very promising experiments.

The sense of smell in fish is a fascinating subject of research throughout the world. Since all life on earth had its origins in the sea, there appears to be no reason to believe that the fish's sense of smell is inferior to that of animals on land. Thus, a sensation was created when a celebrated professor in Munich, Nobel Prize winner Karl v. Frisch, was able to demonstrate how little odour was needed to cause a perch to change its behaviour pattern. Put a few drops of water from a tank in which a pike has been swimming, into a tank containing European perch, and the perch will immediately become frightened. It has smelled its enemy, the pike.

Interesting observations on an international scale are being made at the present time. With the help of electronic devices it is possible to 'read' the reactions of a fish's brain to changes in the odour of the water. It turns out that fish are able to register a scent in water that has been diluted to such a degree that it is beyond our comprehension. Experiments have been carried on at the University of Oslo which prove that a single gramme of a scented substance may be detected by fish in an amount of water that corresponds to 10 million cubic metres of water. One indication that the fish is stimulated is a noticeably more rapid pulse.

The salmon's sense of smell is well known and has been documented, not least by American research. The question of *which* odours it smells is more uncertain. At the University of Oslo in recent years, an astonishing theory has been advanced as to how fish of the salmon family are able to find their way back to the rivers of their childhood. In experiments with sea char, scientist Hans Nordeng asserts he is able to prove that it is not so much the characteristic odour of the river that guides the fish, but that the char smell their kin – the young specimens that still remain in the river. Fish of the salmon family give off a strong odour, pheromones, from the mucous membranes of the skin. It is these pheromones which enable the salmonoids in the sea to find their way back to the correct river.

Since antiquity, literature about fish has been concerned with the fish's sense of smell. Countless 'formulas' exist for scents that attract fish – from cockroaches to cosmetics. Newspapers and weekly magazines are constantly bringing new information from this front! The wisest thing to do is to note the experience of the professional fisherman. The clergyman from Tranöy, who wrote about fishing Greenland shark in Northern Norway in the eighteenth century, noted that the fishermen would not fish for Greenland shark until some groundbait had been dropped well in advance. Indeed, the worse it smelled, the better. The fishermen could tell him that, depending on the current, the voracious Greenland shark could smell the groundbait up to twenty miles away.

A scientist attached to FAO, E. F. Akyüz, recounts his experiments with shark. From the side of the boat he dropped small fish which attracted the sharks. Some 20–40 sharks started a fight for the fishes. He then dropped a baited hook which was taken by one of the sharks. Almost immediately the shark caught on the hook was attacked and badly cut by the other sharks, because, as is known, sharks are inclined to attack wounded creatures. But how could the other sharks know that this one was caught by a hook? Mr Akyüz does not believe that the sharks noticed any difference in the way the caught shark swam – they were all moving in a wild chaotic manner. The reasonable explanation, suggests Mr Akyüz, is that the alarm reaction of the shark caught by the hook probably released through its skin a chemical substance which stimulated the olfactory or some other sense of the other sharks, causing them to attack the vulnerable and injured shark, and which can be likened to the release of sweat in a human being due to a sudden fright.

Sweat contains a specific chemical substance – serin – which has an unusually powerful odour. If it can possibly attract sharks, it may repel other fish. Perhaps it is worthwhile for sports fishermen in lakes and rivers to know that the salmonoids are extremely sensitive to this odour. . . .

Whoever discovers the correct odour for the artificial baits of the future will become very rich indeed.

But how should we view man's ability to preserve the stock of fish? One would have to be quite naïve to regard the coming years in an optimistic light.

With more than 70% of the earth's surface covered by water (not including rivers and lakes) it has been tempting to believe that the waters conceal great reserves of food for the earth's growing population.

But recent studies no longer speak of the 'endless riches' of the sea. In the Norwegian Official Report on Natural Resources, 1972, one learns that the great ocean areas do not produce any more life than the earth's deserts. The oceans comprise 90% of the total marine environment. Thus, it is the land, not the sea, on which the world must pin its hopes. True enough, the yield of the sea is rich in protein, and the most fertile areas of the sea (certain shallower sections of the continental shelves, the sea near the estuaries of rivers, around coral reefs, atolls, etc.) have a production capacity that is comparable to the most highly developed cultures on land. But these most fertile areas of the sea comprise no more than one tenth of the waters.

The future of fishing will entirely depend on mankind. Will everyone do his best to slow down the pollution of the oceans and rivers? And will the responsible leaders in every country realize that it is in their own interests to regulate the catches – not only by entering into agreements, but also by adhering to them?

Given that they will, the experts of the United Nations Food and Agriculture Organization (FAO) are still optimistic. The full potentialities of the fisheries in the southern hemisphere have not yet been exploited and, if nations carry into effect the necessary fishing regulations in every area, the experts believe that the world's supply will not be in danger. On the contrary, it could be increased – perhaps even doubled.

Today there are still enough fish to rebuild the stocks of threatened species. This is also true of the great oceans, where all important forms of life are to be found only in the uppermost layers of the water.

When Thor Heyerdahl and his crew of five left the coast of Peru in 1947 on board the Kon-Tiki raft, they knew there would be enough fish to catch for food in the waters near the coast. They were uncertain, however, of the potential food supply in the open sea. They were relieved and suprised when it turned out that the raft was surrounded by fish at all times. They could choose whatever they wanted for dinner five minutes before the meal.

On all his expeditions Thor Heyerdahl has been well-supplied with Mustad hooks. Only on one occasion were there too few of the stronger types of hooks. So he made his own equipment: using metal wire he fastened three ordinary hooks together into a satisfactory treble hook. A shark bit, and when the powerful creature had been landed, Thor Heyerdahl was so impressed by the strength of each of the three relatively small, hooks, that he immediately sent a congratulatory telegram to the factory from his raft.

Life began in the sea. Will it also end in the sea?

Time alone will show whether the last fish caught by man will be taken on a Mustad fish hook. The firm will continue doing its utmost through research and development to ensure that the fish hook will be as indispensable in the future as it has been in the past – 'this unpretentious implement that has accompanied Man since the beginning of time.'

A good motto. The fishing village of Kilrenny in Scotland has these words of admonition in its coat of arms: Always keep your fish hook out!